W9-CSA-984

Solid State Devices 1986

Solid State Devices 1986

Twelve invited papers presented at the Sixteenth European Solid State Device Research Conference (ESSDERC) held at the University of Cambridge, 8–11 September 1986

Edited by D F Moore

Institute of Physics Conference Series Number 82
Institute of Physics, Bristol and Boston

CODEN IPHSAC 82 1–203 (1987)

British Library Cataloguing in Publication Data

European Solid State Device Research
Conference (*16th: 1986: University of Cambridge*)
Solid state devices 1986: twelve invited papers presented at the Sixteenth European Solid State Device Research Conference (ESSDERC) held at the University of Cambridge, 8–11 September 1986.——
(Conference series, ISSN 0305-2346; no. 82)
1. Semiconductors
I. Title II. Moore, D.F. III. Series
621.3815′2 TK7871.85

ISBN 0-85498-173-X
ISSN 0305-2346

Organising Committee
B L H Wilson OBE (Chairman), P Selway (Vice-Chairman), M Cardwell (Secretary), A N Broers, S Partridge, J E Carroll, D F Moore, W Milne, C A P Foxell, K Board, B Weiss, G Soncini, J P Bailbe

Programme Committee
A N Broers, P Balk, P Scovell, G Jones, D Wolters, C L Claeys, G Soncini, S Partridge, M Roche, J D Speight, H DeGraaf, G Declerck, A Wieder, M Cardwell, M Constant, P Weisglas, M Pilkuhn, M Rocchi, V Ghargia

Honorary Editor
D F Moore

The 16th European Solid State Device Research Conference was organised by The Institute of Physics and co-sponsored by The Institution of Electrical Engineers, The Institution of Electrical and Electronics Engineers, The Institute of Electronic and Radio Engineers and The European Physical Society. The following organisations have made contributions: British Telecom, The Department of Trade and Industry, GEC, Philips, Plessey Research, STL, Thorn EMI, US Army R and D Standardisation and US Airforce Office of Scientific Research.

Published under the Institute of Physics imprint by IOP Publishing Ltd
Techno House, Redcliffe Way, Bristol BS1 6NX, England
PO Box 230, Accord, MA 02018, USA

Printed in Great Britain by J W Arrowsmith Ltd, Bristol

Preface

In September 1986 the annual European Solid State Device Research Conference ESSDERC was held in the Cambridge University Engineering Department, Cambridge, England. During the four days of technical sessions a large volume of novel material was presented both in the invited papers in plenary sessions and in the contributed papers which were timetabled in three parallel sessions. However, in the interests of timely publication of the highlights of the conference, 12 of the invited papers are published in this volume but, unfortunately, none of the 108 contributed papers.

The papers in this book are representative of the main topics of interest at the conference: they are arranged in alphabetic order of first author. A larger number of the papers than usual deal with fabrication technology such as epitaxial materials growth and interfaces, x-ray lithography and dry pattern transfer. This reflects the importance of a close relationship between the device designer and the device technologist in a team developing novel devices.

The papers comprise timely reviews of topics in both silicon and gallium arsenide device technology, but the list is by no means exhaustive and the reader is recommended to refer to ESSDERC proceedings in the recent past for concise reviews of related topics.

At the conference the 300 delegates came from 24 countries, including 87 from the United Kingdom, 66 from West Germany, 34 from France and 24 from The Netherlands. They were comfortably accommodated in Queens' College, and the short walk through Cambridge to the Engineering Department was a good opportunity for delegates to meet informally. The conference reception was held in the University Combination Room, a meeting place since the late fourteenth century, and Professor Pippard emphasised the longstanding relationship between device physicists and device engineers. At the conference dinner in Queens' College the keynote speaker D Roberts, paid tribute to the late D Hooper of GEC Research who contributed so much to the field of solid state devices and to the success of previous ESSDERC conferences.

Finally, we would like to thank all those who contributed to the success of the conference by planning programmes and speakers, chairing sessions and doing the many other organisational tasks. The editor thanks C Barry, R Calder and N Robertson who helped turn camera ready copies of manuscripts into Camera Ready Copies for this book.

D F Moore

Contents

Inst. Phys. Conf. Ser. No. 82
Paper presented at ESSDERC 1986, Cambridge 8–11 Sept. 1986

New frontiers in power devices

B. Jayant Baliga
Coolidge Fellow and Manager, High Voltage Device Program
General Electric Company
Corporate Research and Development Center
Schenectady, NY 12301, U.S.A.

Abstract. The advent of MOS-controlled power devices is setting the stage for a revolutionary advance in power semiconductor technology. This paper reviews the development of the two key devices—the power MOSFET and the IGT—which have been responsible for the resurgence of power electronics. These devices offer a high input impedance feature which enables the integration of their gate drive circuit by using high voltage IC technology. This combination of MOS controlled power devices and HVICs is resulting in a reduction in power system cost by an order of magnitude creating the opportunity for power electronics to penetrate cost sensitive consumer and industrial electronic systems.

1. Introduction

Until the 1970s, the power industry relied upon bipolar power semiconductor devices. Among this class of devices, the power thyristor has been used in high power systems such as high voltage DC transmission networks. A steady growth in the ratings of these devices has occurred over the last 30 years. At present, devices capable of handling 6500 volts and 1000 amperes are being manufactured out of 77 mm diameter wafers. Innovations in this area are directed towards improved edge termination techniques to raise breakdown voltage and creating optical triggering capability to simplify device isolation in the high voltage networks.

For higher frequency switching systems, the bipolar transistor has been extensively used. Its ratings have grown steadily over the years and devices capable of handling several hundred amperes at 500 volts are available today with turn-off times of less than one microsecond. The most important problem that has been encountered with the application of these devices is associated with the large input current required to control these devices. Even with the development of monolithic Darlington structures, the current gain of the bipolar transistor is less than 10. This results in the need to deliver many amperes of current to control the device. Such high currents cannot be provided from an integrated circuit. The discrete part control circuit for the bipolar transistor is bulky and expensive.

The power MOS technology reviewed in this paper was developed in response to the need to create a power device structure that can be controlled by using very low currents. This feature, in conjunction with the creation of high voltage integrated circuits (HVICs), allows the integration of their control circuitry. The development of the power MOS devices and HVICs is resulting in a reduction in the cost of power electronics by an order of magnitude. This not only impacts existing power systems but creates the opportunity to introduce power electronics into new cost-sensitive consumer and industrial systems.

2. Power MOSFET

The evolution of power MOSFET technology can be traced to MOS-IC technology. The first attempt at combining high voltage with MOS structures was performed in lateral devices which could be integrated with control circuitry for driving transducers (Plummer and Meindl 1976). As the advantages of the high input impedance of a power MOS device became apparent, vertical structures were developed as high power discrete transistors. In the early 1970s, there were two technologies - VMOS and DMOS - with strong advocates for both. Over the years, many disadvantages for the VMOS structure were discovered. This has led to its obsolescence. At present, all commercial devices are being made using the DMOS structure.

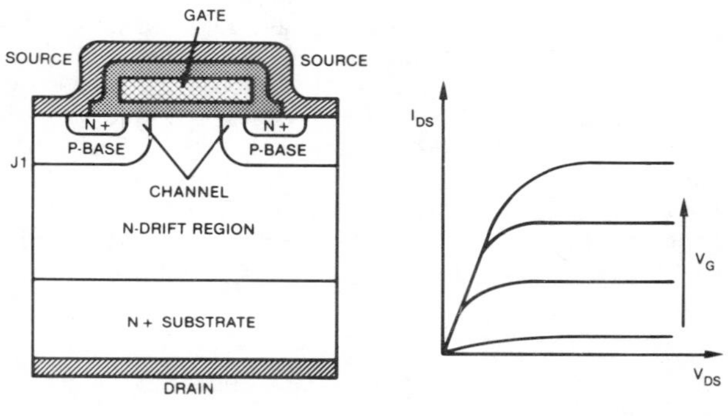

Fig. 1 Power MOSFET cross-section and its output characteristics.

The basic vertical DMOS structure is illustrated in Fig. 1. The device contains two important regions--the N-drift region that supports the high voltage and the channel region which controls the on/off state of the device. When the gate is connected to the source, the device blocks current flow for positive drain voltages. The P-N junction (J1) becomes reversed biased and a depletion layer extends into the N-drift region. The resistivity and thickness of this region determine the breakdown voltage. When a positive gate bias is applied, an n-channel (inversion layer) is formed at the surface of the P-base region. This allows electron transport from the N^+ emitter to the N^+ drain (substrate). Since increasing the gate voltage enhances the channel conductivity, the device current increases with increasing gate voltage as illustrated on the right hand side in Fig. 1.

2.1 On-Resistance

The maximum current handling capability of the power MOSFET is determined by its on-resistance. The on-resistance is defined as the resistance of the device at low drain currents. It is primarily determined by a combination of the resistances of the channel and drift region. Other components that can affect the device are the contact and substrate resistances.

The channel and drift region resistance are in turn determined by the DMOS cell design. The cell design consists of many variables: (1) cell topology (circular, hexagonal, square, linear, etc.); (2) the polysilicon width; (3) the polysilicon window; (4) the channel length (determined by the difference in lateral diffusion between the P-base and the N^+ emitter); (5) gate oxide thickness; and (6) the gate drive voltage. The detailed design of the DMOS cell is beyond the scope of this paper and has been treated in depth elsewhere (Baliga and Chen 1984). An example of the impact of changing the polysilicon width upon the specific on-resistance (i.e. on-resistance for unit area) is provided in Fig. 2 for the case of a square cell design. In this figure, the impact of changing the cell window size is also shown. It is apparent that the on-resistance can be greatly reduced by decreasing the cell window size.

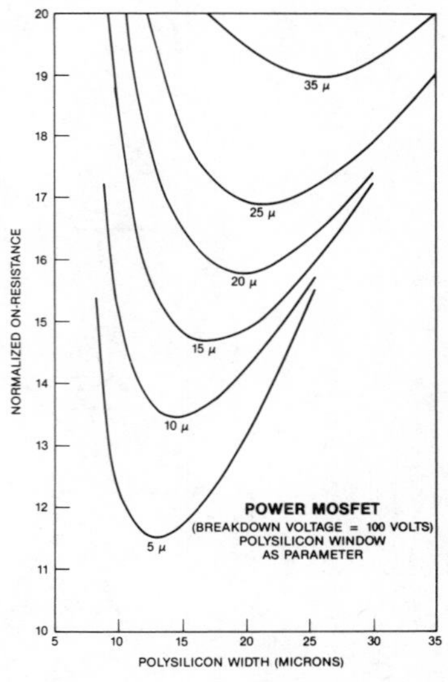

Fig. 2 Impact of changing the polysilicon (gate overlap) width upon the on-resistance of a power MOSFET. Note the improvement in on-resistance as the polysilicon window is reduced from 35 microns to 5 microns.

The minimum cell window size is determined by the device process technology, particularly the lithographic tolerances used in fabricating the cell structure. Over the years, there has been a continual advancement in this area resulting in the ability to reduce the cell size while maintaining a reasonable yield. The progressive decrease in cell size is illustrated in Fig. 3. At the start of this decade, the minimum cell size was about 50 microns. This corresponds to the upper curves in Fig. 2 which represent an optimum cell design with a cell window of 25 to 35 microns and a polysilicon width of 20 to 25 microns. With improvements in process technology (such as projection printing, better step coverage, dry etching, etc.), the minimum DMOS cell size has now been reduced to 20 microns. This corresponds to the lower curves in Fig. 2. As a consequence of this, the minimum specific on-resistance achievable for the DMOSFET has been reduced from about 7 milliohm-cm^2 to 1.7 milliohm-cm^2 as illustrated in Fig. 4. Simultaneously, the yield of larger chips has been improving and the largest power MOSFET chips that are manufacturable with reasonable yield have grown from 2.5 mm in 1980 to 6.4 mm in 1986 (see Fig. 3).

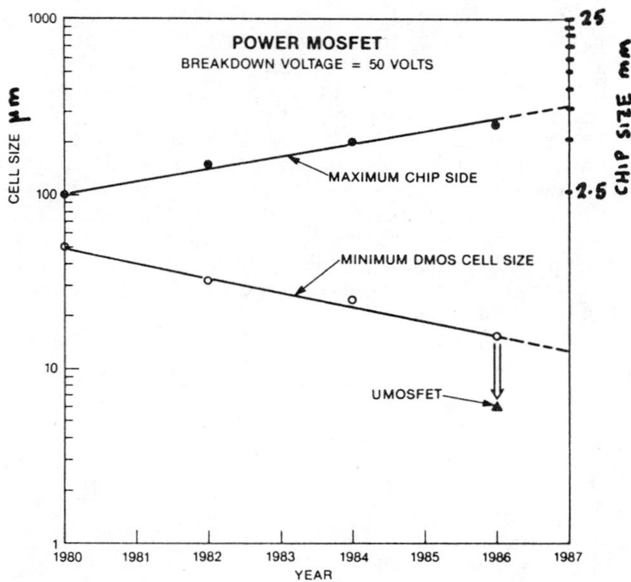

Fig. 3 Trends in the reduction of DMOS cell size and increase in maximum chip size.

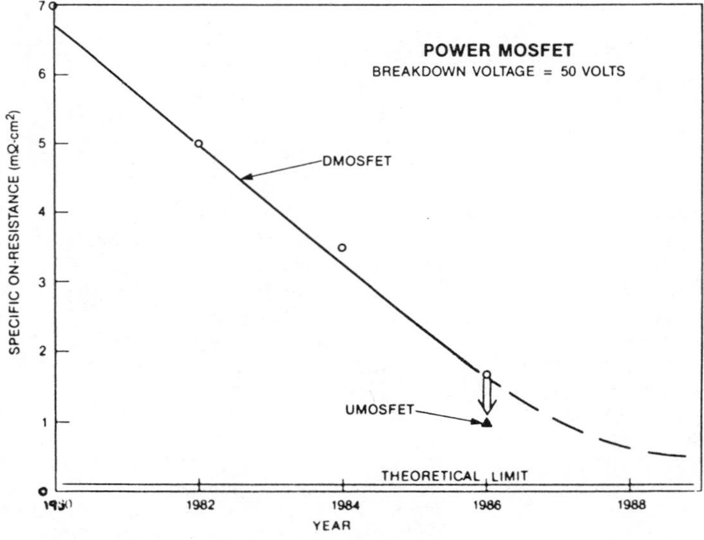

Fig. 4 Trends in the reduction of the specific on-resistance for 50 volt power MOSFETs.

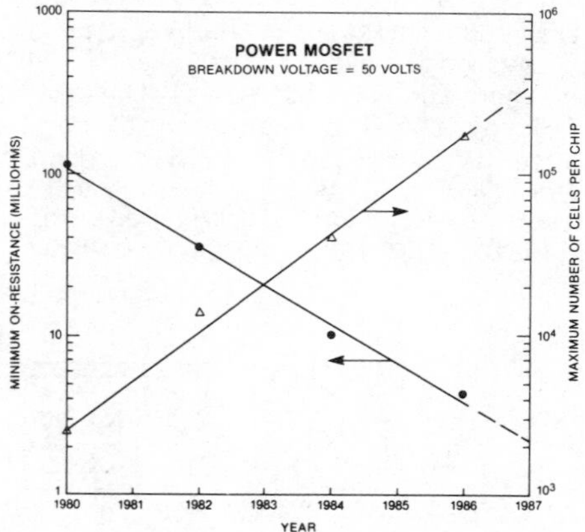

Fig. 5 Growth of the number of cells per chip for power MOSFETs and the resulting reduction in the lowest available on-resistance.

The overall impact of the reduction in cell size and increase in the chip size is a rapid growth in maximum number of cells that can be integrated on a single chip. It can be seen from Fig. 5, that starting with only 2500 cells per chip in 1980, the power MOS industry now has the capability to integrate nearly 200,000 cells per chip. This ability has resulted in a dramatic reduction in the lowest on-resistance available in a power MOSFET. Early devices (circa 1980) had on-resistances of over 100 milliohms. Today, devices with an on-resistance of less than 5 milliohms are becoming available. The resulting steady growth in power ratings of the DMOSFET implies greater penetration of this technology in systems.

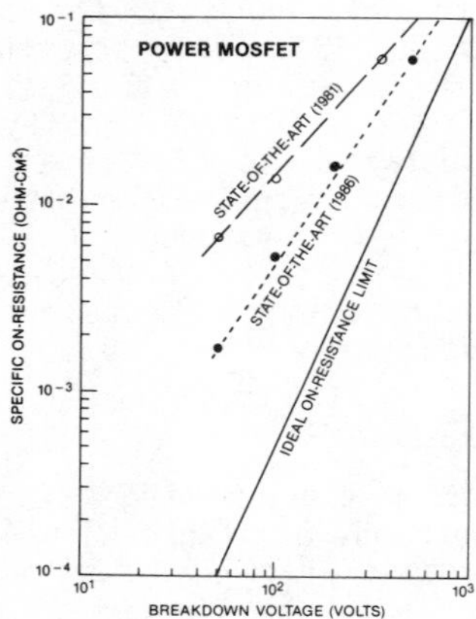

Fig. 6 Improvement in specific on-resistance for power MOSFETs compared with the ideal limit for silicon unipolar devices.

It is worth pointing out that improvements discussed above for the low voltage power MOSFET are also applicable at higher voltages. A comparison of the specific on-resistance achievable in 1981 (Baliga 1981) with that obtained today in state-of-the-art devices is given in Fig. 6. In this figure, the ideal on-resistance limit for silicon is also provided because it defines the limit for this technology. An important conclusion that can be derived from this figure is that there is still much room for improvement of the specific on-resistance especially for the low voltage devices.

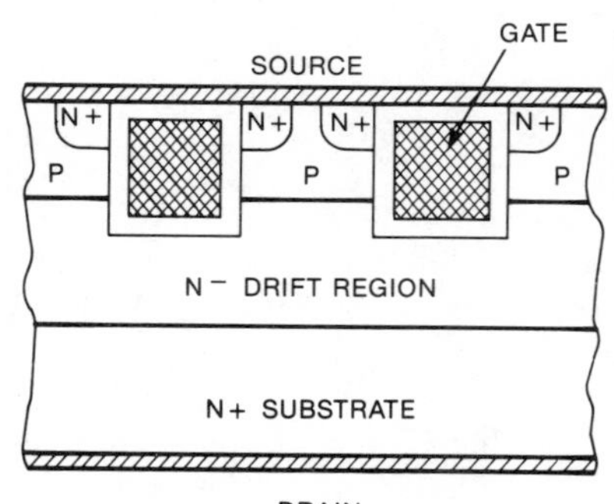

Fig. 7 A cross-section of a vertical channel MOSFET structure fabricated using trench gate technology.

2.2 UMOSFET Structure

The above discussion was confined to the DMOS cell structure. Some inherent disadvantages of that structure are the parasitic resistance of the region between the P-base diffusions (often called the JFET region) and the low channel density (channel width per cm^2 of active area). A big improvement in on-resistance can be obtained by using the UMOSFET structure illustrated in Fig. 7 (Ueda et al 1985, Chang et al 1986). This structure is fabricated by forming vertically walled trenches which are refilled with the polysilicon gate after formation of the gate oxide. With this structure, the JFET region is eliminated. Further, a much smaller cell size can be achieved which results in an increase in the channel density. The impact of the UMOSFET structure is illustrated in Fig. 3 and 4. By using a cell size of 6 microns, a specific on-resistance of 1 milliohm-cm^2 has been achieved (Chang et al 1986). Although this technology is in a developmental phase and no commercial devices are as yet available, it represents an important innovation that could drive the power MOSFET ratings upwards in the future.

3. Insulated Gate Transistor

Although power MOSFETs offer the advantage of a low control current, the sharp increase in their on-resistance with increasing voltage (see Fig. 6) results in poor power handling capability compared with the bipolar transistor. The impact of the power MOSFET is strongest in systems operating at below 200 volts. A prime example of this is automotive electronics. To impact higher voltage systems, a new class of power devices based upon a combination of bipolar and MOS physics has been developed. The most mature example among these devices is the insulated-gate-transistor (IGT) (Baliga et al 1984).

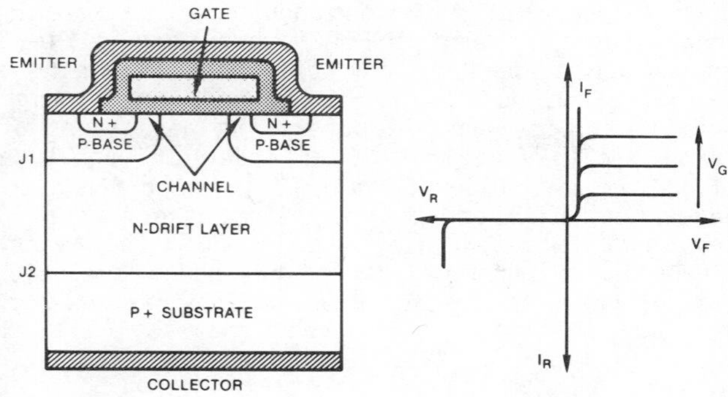

Fig. 8 Cross-section and output characteristics of the Insulated Gate Transistor.

The IGT structure is illustrated in Fig. 8. This structure is similar to a power MOSFET with the exception of an additional junction (J2) at the bottom of the wafer. A detailed description of device operation has been provided elsewhere (Baliga et al 1984). The most important feature of the IGT is the reduction in the series resistance of the drift layer due to minority carrier injection from the P^+ substrate during forward conduction. This results in a rectifier-like forward conduction characteristic.

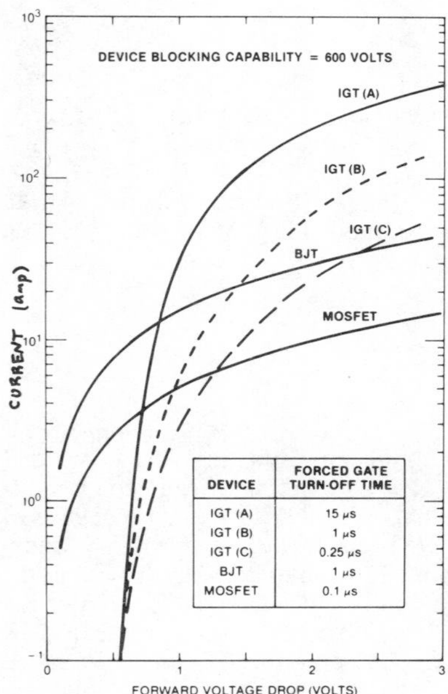

Fig. 9 Comparison for the forward conduction characteristics of IGTs with various turn-off speeds with a power MOSFET and bipolar transistor.

A comparison of the IGT forward conduction characteristics with the bipolar transistor and power MOSFET is provided in Fig. 9. The three IGT curves represent devices with different switching speeds (Baliga 1983). From Fig. 9, it can be concluded that for high voltage, low frequency applications, the IGT offers the advantages of a high input impedance and a chip size that is an order of magnitude smaller than the MOSFET. This advantage of the IGT has been found to be applicable for frequencies as high as 50 kHz. Since the process technology for IGTs closely parallels that for power MOSFET, the ability to increase the power handling capability per cm^2 has had a dramatic impact on the availability of power MOS gated chips with higher power ratings.

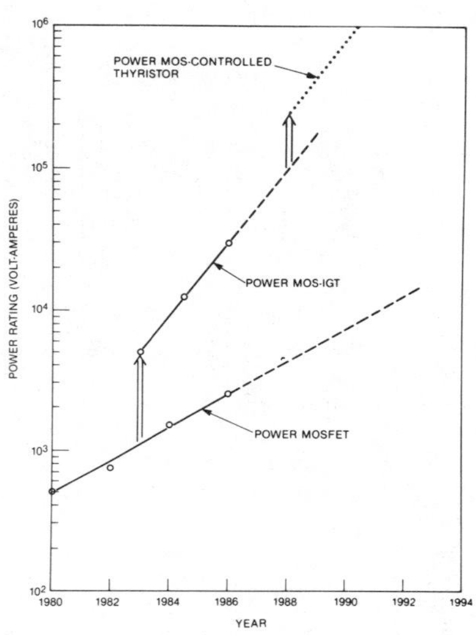

Fig. 10 Increase in the power handling capability of power MOS technology.

4. Trends

Since the penetration of power MOS technology into an electronic system is dependent upon the availability of chips with higher power handling capability, it is interesting to track the growth in power ratings over the years. In the early 1980s, the only power MOS technology that was available was the MOSFET technology. As discussed earlier, the power handling capability of these devices is dependent upon the ability to reduce the specific on-resistance and to increase the maximum die size. From the data shown in Fig. 3 and 4, it can be concluded that: (1) the power MOS chip area is doubling every two years; (2) the specific on-resistance of power MOSFETs is decreasing by a factor of 1.4 every two years; (3) the power MOSFET on-resistance is decreasing by a factor of 3 every two years. These trends are expected to continue.

The cumulative impact of these developments upon the power handling capability of power MOSFETs is shown in Fig. 10. The power ratings have grown from 500 watts in 1980 to 2500 watts today. The impact of the introduction of the IGT in 1983 can also be clearly seen in this figure. The power handling capability jumped by a factor of 5 at that time. Today IGTs are available with ratings of 1200 volts and 25 amperes. The IGT ratings are, therefore, growing at a much faster pace than for power MOSFETs. This is a direct consequence of the ease with which IGT voltage ratings can be increased compared with a power MOSFET. In the IGT, the current handling capability per unit area decreases as the square root of the voltage rating compared with a reduction proportional to the 2.5 power of the voltage for the power MOSFET (Chow et al 1985). Based upon the data in Fig. 10, the IGT ratings can be projected to grow by a factor of 3.4 every two years. This is twice as fast as the rate of increase in power MOSFET ratings.

The innovations made in power structures that have led to the observed rapid growth in power ratings are expected to continue. Another MOS-bipolar device that holds great promise is the MOS gated thyristor. The first MOS gated thyristors were reported in 1979 (Baliga 1979). These devices had the advantages of a MOS controlled turn-on with improved di/dt and dv/dt capability. More recently, the concept has been extended to achieve MOS gated turn-off (Temple 1984, Stoisiek and Strack 1985). These devices operate like thyristors in the on-state. This results in even better modulation of the drift layer resistance compared with the IGT. An increase in power handling capability per unit chip area by a factor of 2 to 3 over the IGT can be projected as illustrated in Fig. 10. The commercial introduction of these devices in the future can be expected to result in another boost to the power handling capability of the power MOS technology.

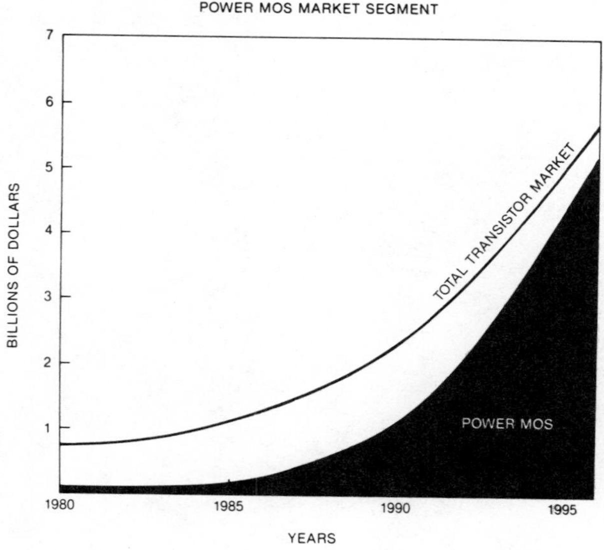

Fig. 11 Trends in the total transistor market. Note that an increasing fraction of the market will be taken over by the power MOS technology.

As the ratings of power MOS devices continues to grow and approaches that of a larger and larger segment of the bipolar transistor ratings, a continual erosion of the bipolar transistor market can be foreseen. It has been projected that with the increasing application of power electronics in motor drives and factory automation, the power transistor market will grow from its present level of about $1 Billion to $5 Billion in 1995. As illustrated in Fig. 11, at present the market share for power MOS transistors represents about 10 percent of the total transistor market. However, by the mid 1990s, the power MOS transistor segment will grow to over 80 percent of the total market and the power MOS transistor technology can be expected to become the dominant discrete device technology for the 1990s.

References

Baliga B. J. 1979 Electronics Letters 15,645

Baliga B. J. 1981 IEEE Spectrum 18 (12), 42

Baliga B. J. 1983 IEEE Electron Device Letters EDL-4,42

Baliga B. J., Adler M. S., Love R. P., Gray P. V., and Zommer N. D. 1984 IEEE Trans Electron Devices ED-31,821

Baliga B. J. and Chen D. Y. 1984 "Power Transistors" (IEEE Press Book)

Chang H. R., Black R. D., Temple V. A.K., Tantraporn W. and Baliga B. J. 1986 Int. Electron Devices Meeting

Chow T. P. and Baliga B. J. 1985 IEEE Electron Device Letters EDL-6,161

Plummer J. D. and Meindl J. D. 1976 IEEE Trans. Electron Devices ED-21,778

Stoisiek M. and Strack H. 1985 Int. Electron Devices Meeting 158

Temple V. A. K. 1984 Int. Electron Devices Meeting 282

Ueda D., Takagi H. and Kano G. 1985 IEEE Trans. Electron Devices ED-32,2

Inst. Phys. Conf. Ser. No. 82
Paper presented at ESSDERC 1986, Cambridge 8–11 Sept. 1986

The device application of silicon molecular beam epitaxy

John C. Bean

AT&T Bell Laboratories, Murray Hill, NJ 07974 USA

Abstract. This paper will discuss the device potential of materials grown epitaxially on silicon by the use of molecular beam epitaxy. The following application categories are discussed: Homoepitaxial structures incorporating very thin silicon layers or layers with hyperabrupt doping profiles. Epitaxial metal devices such as the metal base and permeable base transistor. Devices employing epitaxial insulators as gate and isolation layers. Semiconductor heterostructures exploiting Ge_xSi_{1-x}/Si strained layer epitaxy.

In the mid 1970's, silicon molecular beam epitaxy (Si-MBE) first produced material of high crystallographic quality.[1,2] This material was rapidly applied in a number of devices that served to further evaluate material quality but did not demonstrate a uniquely desirable MBE capability.[1,3,4,5,6,7,8] This situation is now changing. More sophisticated MBE capabilities, largely in the area of heteroepitaxy, have now produced device performance beyond the conventional technology and this has led to the first commercial application of the technique. This paper will briefly review that recent device work.[9,10]

Homoepitaxy and Hyperabrupt Doping

The original work on Si-MBE was directed towards applications requiring either very thin silicon layers or layers with abrupt, complex doping profiles. Early goals included IMPATT, Gunn and Read microwave diodes. Conventional fabrication of a double drift Read diode with a $p^+-\pi-p^+-n^+-\nu-n^+$ doping profile could involve growth of π epi on a p^+ substrate followed by p^+ and n^+ ion implantations followed by a ν epi layer followed by a third n^+ implant. Because Si-MBE can be done well below 800°C, diffusion and autodoping are avoided and such a profile can in principle be grown in a single epi growth step.

As described recently by Luy et.al.,[11] high performance silicon microwave diodes have been fabricated by Si-MBE and are being offered by AEG Corp. as a commercial product.[12] MBE is used to grow n^+ and n layers on an n^+ substrate with respective layer thicknesses of 0.8 and 0.3 microns. A boron contact is then diffused, the substrate thinned and a mesa defined by lithography to produce a single drift region diode. At 101 GHz, these devices produce a CW power output of 250 mW with an efficiency of 5.8% at a current density of 45 kA/cm^2.

A second and potentially more important application of Si-MBE is in the replacement of chemically vapor deposited (CVD) epitaxy for growth of thin undoped or lightly doped layers in IC fabrication. These layers frequently overlay heavily doped Sb regions used to minimize lateral resistance. CVD growth temperatures can produce outdiffusion with lateral non-uniformities requiring the use of exaggerated epi layer thicknesses. These problems are avoided in MBE and epi thickness can be tailored to produce maximum circuit performance.

This tailoring process has been exploited by Kasper and Worner in the fabrication of the high speed frequency divider circuit shown in Fig. 1.[13,14] The circuit, using master-slave flip-flops in emitter coupled logic, was originally designed for CVD epi to produce operation at 900 MHz. The substitution of MBE permitted a reduction of epi thickness and vertical device dimensions. Together, these reductions produced an increase of operating frequency to 2.8 GHz. More significantly, with the exception of the MBE step, the entire circuit was fabricated on the conventional integrated circuit line and MBE based circuits were reported to have a yield comparable to those based on the much more mature CVD process.

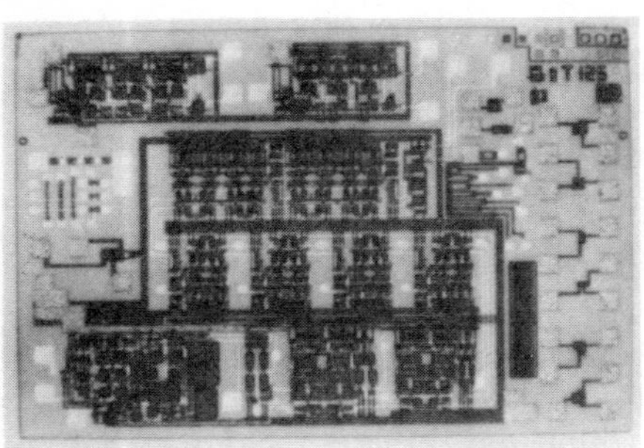

Fig. 1

Photograph of Si-MBE based ECL frequency divider fabricated by AEG Corp.

Epitaxial Metal Devices

The metals $NiSi_2$ and $CoSi_2$ have a cubic lattice structure with lattice spacings closely matched to that of silicon. They can be grown in single crystal form on silicon either by codeposition of metal and silicon[15] or by deposition of pure metal and subsequent thermal reaction.[16] The resultant crystal can directly extend the symmetry of the silicon substrate or involve a 180° rotation. The rotational alignment, surface and interface morphology, and pinhole density can be effected and to some extent controlled by depositing atomic monolayer dimension metal "template" layers prior to growth of the thick silicide layers.[17,18]

Epitaxial metal silicide layers have been used by Rosencher et.al.[19] and Hensel et.al.[20,21,22] as base layers in the successful fabrication of transistors. Common base current gains of 0.01-0.95 are reported and voltage gains of over ten have been produced at 100 kHz. At the outset both groups sought classic metal base transistor action.[23] It is not yet certain if this was achieved. The epitaxial metal silicides have a peculiar tendency of retaining small pinholes that fill with silicon when overgrown. This produces a parasitic permeable base transistor[24] as indicated in Fig. 2. Rosencher et.al.[19] report two distinct transistor I-V characteristics, dependent on metal silicide growth conditions, which are interpreted as either metal or permeable base action.[25] Hensel counters that computer simulations show both characteristics for permeable base action and that observed current gains decrease systematically from 0.95 to 0.01 as pinhole density decreases.

Permeable transistor action may in fact be quite desirable especially when one considers that these naturally occurring pinholes eliminate the need for the sophisticated electron beam lithography used in previous transistors of this type. Simulations suggest that metal silicide permeable base transistors may be capable of current gains of over 100 and operating frequencies above 20 GHz.[26]

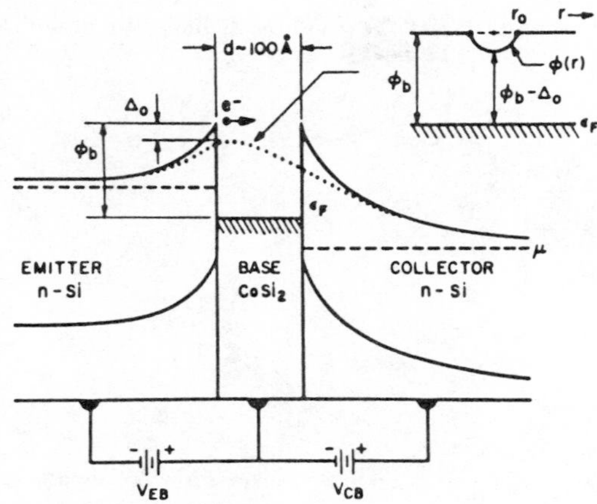

Fig. 2

Schematic diagram of potential barriers produced by metal base and permeable base action in epitaxial metal transistors.

Epitaxial Insulator Devices

As in the metal silicide case, the insulator calcium fluoride is closely lattice matched to silicon and can be grown epitaxially in single crystal form by MBE.[27,28] Growth is greatly simplified by the fact that CaF_2 evaporates as a molecule (i.e. congruently) and thus requires only a single evaporation source to produce stoichiometric growth. CaF_2 has a melting point of 1360°C and in single crystal form it is reported to be essentially insoluble in water.

Smith et.al.[29,30] have used CaF_2 as the gate layer in a metal epitaxial insulator field effect transistor (MEISFET). The enhancement mode n-channel device yielded a peak room temperature channel mobility of 400 cm^2/V-sec., and dielectric breakdown voltages as high as $3x10^6$V/cm. Capacitance measurements indicated a static dielectric constant of 6.8 and interface state densities as low as $5x10^{11}/cm^2$-eV. In simpler diode structures, interface state densities of $7x10^{10}/cm^2$-eV were reported.[31] This interface state density is remarkably low considering the relatively undeveloped state of this material and the physics-style fabrication process.

The natural extension of the above experiments is to use epitaxial CaF_2 for vertical isolation in a three dimensional device structure. Asano et.al.[32] have used this approach to fabricate a single crystal N-MOSFET on CaF_2 on Si. The device exhibited a channel mobility of 580 cm^2/V-sec., or about 80% of the value observed in a control device fabricated in bulk Si. The degradation was apparently due to residual defects in the Si on CaF_2 that have proved difficult to eliminate using conventional MBE growth procedures. Recent work by Pfeiffer and Phillips et.al.[33,34,35] may provide a solution to this problem. By applying a post growth rapid thermal anneal of 20 sec. at 1100°C, they produce a dramatic reduction in CaF_2 defect density. If this improvement results in an increase of Si overlayer quality, bulk device characteristics may be possible.

Ge_xSi_{1-x}*/Si Heterostructure Devices*

Work on silicon based semiconductor heterostructures had been frustrated by the absence of a lattice matched column IV semiconductor. This situation has been overcome with the discovery that Ge and Ge_xSi_{1-x} alloys can grow without defects by means of "strained layer epitaxy" on silicon substrates.[36,37] This mode of growth can be maintained to thicknesses much larger than predicted by theory[38,39] or observed in earlier studies[40,41] (Fig. 3).

In addition to eliminating defects, it has been found that strain causes a dramatic reduction of the Ge_xSi_{1-x} bandgap, as shown in Fig. 4. For a heterostructure application calling for a given bandgap difference, this alteration means that less Ge is required. Thicker strained layers can thus be grown or layer thickness can be maintained and Ge fraction reduced to minimize strain.

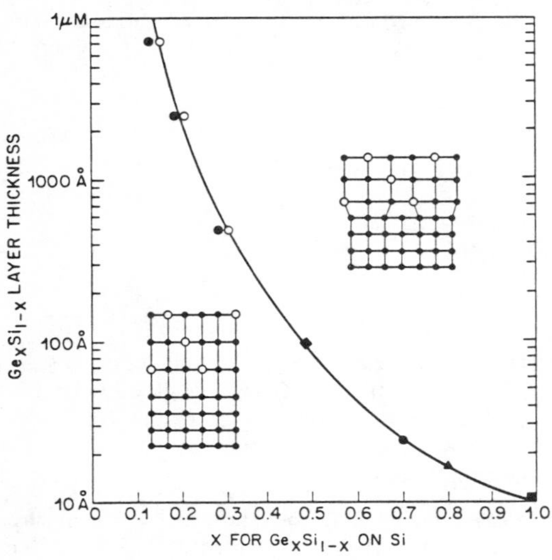

Fig. 3

Maximum thicknesses for defect-free strained layer growth of Ge_xSi_{1-x} on Si.

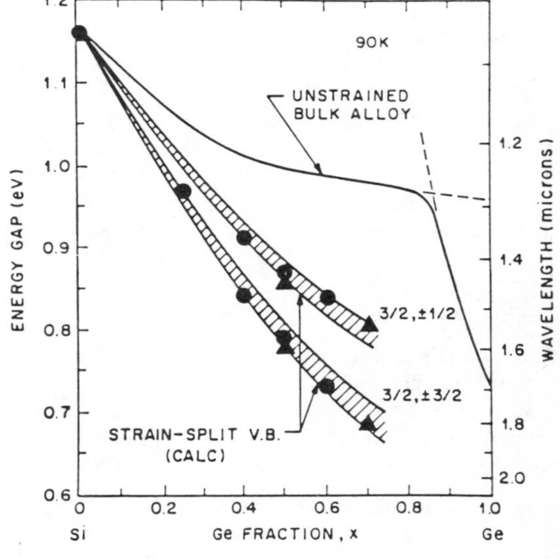

Fig. 4

Calculated and measured values of strained Ge_xSi_{1-x} on Si bandgap and comparison with bulk alloys.

The line up of Ge_xSi_{1-x} and Si band edges has been determined by both p[42,43,44] and n[45,46] modulation doping experiments and by theory.[47] For the simpler case of Ge_xSi_{1-x} strained to conform with an unstrained Si substrate, the alloy band gap falls within the Si bandgap. The bandgap difference is split about 80/20 between the valence and conduction band discontinuities, respectively. If, on the other hand Ge_xSi_{1-x} is graded over a large enough thickness to produce a lattice constant larger than Si, overgrowth with Ge_xSi_{1-x}/Si will produce strain in both materials. This can produce a staggered band line up with the alloy conduction edge above that of Si.

Ge_xSi_{1-x}/Si strained layer epitaxy was first exploited in the successful fabrication of silicon based p–MODFETs[48,49] and n–MODFETs.[50] The p-MODFET employed conventional planar Si IC processing technology including lithography, metallization, ion implantation, activation anneals and thermal oxidation. Modulation doping was maintained despite the metastable layer strain and preliminary structures had a 70 mA/mm channel conductance and a transconductance of ~10 mS/mm. Both enhancement and depletion mode operation were demonstrated. Although the n-MODFET used a less conventional fabrication procedure it achieved a transconductance of 40 mS/mm (with a 1.6μm gate length). Most significantly, this transistor demonstrated a room temperature channel mobility of 1500 cm^2/V-sec.

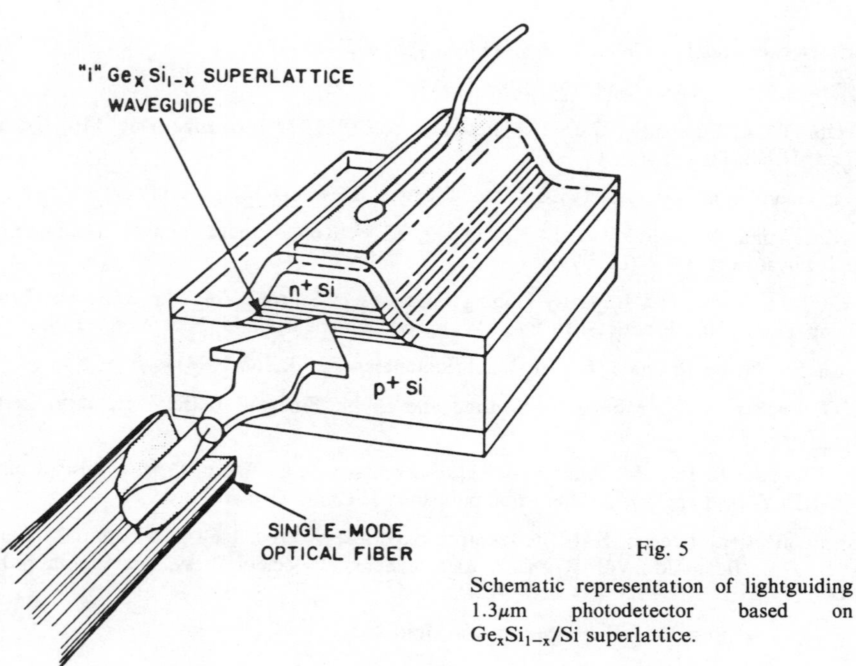

Fig. 5

Schematic representation of lightguiding 1.3μm photodetector based on Ge_xSi_{1-x}/Si superlattice.

Strain reduces the Ge_xSi_{1-x} bandgap to the point where absorption will occur in the 1.3–1.5μm range of critical importance to fiber optic communications. This has been exploited in PIN and APD photodetectors such as that shown in Fig. 5. These structures employ a light guiding geometry to achieve strong optical absorption in a narrow diode structure. In a simple cleaved PIN configuration[51,52] uniform breakdowns of up to 38 V were measured with reverse leakages of $<1\text{x}10^{-3}\text{A/cm}^2$ at -10 V. At 1.3μm, internal quantum efficiencies of over 50 % were measured. With the addition of an avalanche multiplication layer,[53,54,55] APD gains of up to 50 have been produced. Gain bandwidth products of over 48 GHz have been achieved and detectors have sensed 1.275μm optical signals over a 45 km fiber link at 800 MHz with bit error rates of $<10^{-9}$.

Summary

Si-MBE can now produce unique and desirable device structures exploiting mobility enhancements, long wavelength absorption, epitaxial metal conductivity and single crystal dielectric isolation. Fortuitously, these innovations come at a time when there is a general move towards low temperature silicon processing. These reduced temperatures will preserve the unique characteristics of MBE devices and should permit their direct integration with existing silicon circuit technology.

REFERENCES

[1] G. E. Becker and J. C. Bean, J. Appl. Phys. 48, 3395 (1977)

[2] Y. Ota, J. Electrochem. Soc. 124, 1795 (1977)

[3] Y. Ota, W. L. Buchanan, and O. G. Peterson, p. 375, 1977 Int. Elec. Dev. Mtg. Tech. Digest, IEEE, New York

[4] C. A. Goodwin and Y. Ota, IEEE Trans. Electron. Dev. ED-26, 1796 (1979)

[5] Y. Katayama, Y. Shiraki, K. L. Kobayashi, K. F. Komatsubara, and N. Hashimoto, Appl. Phys. Lett. 34, 740 (1979)

[6] J. C. Bean, Chap. 4 in Impurity Doping Processes in Silicon, Vol. 2 in Materials Processing Theory and Practices, F. F. Y. Wang Ed., North Holland, New York (1981)

[7] E. Kasper, H. Barth and J. Freyer, Nachrichtentech Z. 34, 768 (1981)

[8] R. G. Swartz, J. H. McFee, P. Grabbe and S. N. Finegan, IEEE Elec. Dev. Lett. EDL-2, 293 (1981)

[9] For a broader review of current Si-MBE research see: J. C. Bean, Proc. 4th Int. Conf. on MBE, York, Sept 7-10, 1986, to be published J. Cryst. Growth

[10] For a collection of recent Si-MBE research papers see: Proc. 1st Int. Symp. on Silicon MBE, J. C. Bean Ed., Vol. PV 85-7, Electrochemical Society Press, Pennington N.J., (1985)

[11] J. F. Luy, W. Behr, and E. Kasper, p. 236, Ibid.

[12] Private communication, E. Kasper

[13] E. Kasper and K. Worner, p. 429, VLSI Science and Technology, Vol. PV 84-7, Electrochemical Soc, Pennington NJ (1984)

[14] E. Kasper and K. Worner, J. Electrochemical Soc. 132, 2481 (1985)

[15] J. C. Bean and J. M. Poate, Appl. Phys. Lett. 37, 643 (1980)

[16] S. Saitoh, H. Ishiwara and S. Furukawa, Appl. Phys. Lett. 37, 203 (1980)

[17] R. Tung, J. M. Gibson, and J. M. Poate, Phys. Rev. Lett. 50, 429 (1983)

[18] For a recent review see R. Tung, Chap. 10 in Silicon Molecular Beam Epitaxy, E. Kasper and J. C. Bean Eds., CRC Press

[19] E. Rosencher, S. Delage, Y. Camidelli and F. Arnaud d'Avitaya, Electronics Lett. 20, 762 (1984)

[20] J. C. Hensel, A. F. J. Levi, R. T. Tung and J. M. Gibson, Appl. Phys. Lett. 47, 151 (1985)

[21] J. C. Hensel, p. 499, Interfaces and Phenomena, R. H. Nemanich, P. S. Ho and S. S. Lau Eds., Vol. 54, Materials Res. Soc. Conf. Proc., MRS, Pittsburgh (1985)

[22] R. T. Tung, A. F. J. Levi, and J. M. Gibson, Appl. Phys. Lett. 48, 635 (1986)

[23] For reviews see S. M. Sze and H. K Gummel, Sol. St. Elec. 9, 751 (1966) and C. R. Crowell and S. M. Sze, Physics of Solid Films 4, 325 (1967)

[24] See for example C. O. Bozler, G. D. Alley, R. A. Murphy, D. C. Flanders and W. T. Lindley, p. 384, 1979 Int. Elec. Dev. Mtg. Tech Digest, IEEE, New York

[25] F. Arnaud d'Avitaya, private communication

[26] J. Hensel, Appl. Phys. Lett., 49, 522 (1986)

[27] R. F. C. Farrow, P. W. Sullivan, G. M. Williams, G. R. Jones, and C. Cameron, J. Vac. Sci. Technol. 19, 415 (1981)

[28] H. Ishiwara and T. Asano, Appl. Phys. Lett. 40, 66 (1982)

[29] T. P. Smith, J. M. Phillips, W. M. Augustyniak and P. J. Stiles, Appl. Phys. Lett. 45, 907 (1984)

[30] T. P. Smith, J. M. Phillips, R. People, J. M. Gibson and P. J. Stiles, p. 163, Layered Structures, Epitaxy and Interfaces, J. M. Gibson and L. R. Dawson Eds., Proc. Vol. 37, Materials Research Society, Pittsburgh (1985)

[31] R. People, T. P. Smith, J. M. Phillips, W. M. Augustyniak and K. W. Wecht, Ibid. p. 169

[32] T. Asano, Y. Kuriyama and H. Ishiwara, Electronics Lett. 21, 386 (1985)

[33] L. Pfeiffer, J. M. Phillips, T. P. Smith, W. M. Augustyniak and K. W. West, Appl. Phys. Lett. 46, 947 (1985)

[34] J. M. Phillips, L. Pfeiffer, D. C. Joy, T. P. Smith, J. M. Gibson, W. M. Augustyniak, and K. W. West, J. Electrochem. Soc. 133, 224 (1986)

[35] J. M. Phillips and W. M. Augustiniak, Appl. Phys. Lett. 48, 463 (1986)

[36] J. C. Bean, T. T. Sheng, L. C. Feldman, A. T. Fiory and R. T. Lynch, Appl. Phys. Lett. 44, 102 (1984)

[37] J. C. Bean, L. C. Feldman, A. T. Fiory, S. Nakahara and I. K. Robinson, J. Vac. Sci. Technol. A2, 436 (1984)

[38] F. C. Frank and J. H. van der Merwe, Proc. Roy. Soc. A198, 205 (1949) and A198, 216 (1949) and A200, 125 (1949)

[39] J. W. Matthews, J. Vac. Sci. Technol. 12, 126 (1975) and references therein

[40] E. Kasper, H. J. Herzog and H. Kibbel, Appl. Phys. 8, 199 (1975)

[41] E. Kasper and H. J. Herzog, Thin Sol. Films 44, 357 (1977)

[42] R. People, J. C. Bean, D. V. Lang, A. M. Sergent, H. L. Stormer, K. W. Wecht, R. T. Lynch and K. Baldwin, Appl. Phys. Lett. 45, 1231 (1984)

[43] R. People, J. C. Bean and D. V. Lang, J. Vac. Sci. Technol. A3, 846 (1985)

[44] R. People, J. C. Bean and D. V. Lang, p. 360, Proc. 1st Int. Symp. on Silicon MBE, J. C. Bean Ed., Vol. PV 85-7, Electrochemical Society Press, Pennington N.J., (1985)

[45] H. Jorke and H. J. Herzog, Ibid. p. 352

[46] G. Abstreiter, H. Brugger, T. Wolf, H. Jorke and H. J. Herzog, Phys. Rev. Lett. 54, 2441 (1985)

[47] R. People and J. C. Bean, Appl. Phys. Lett. 48, 538 (1986)

[48] T. P. Pearsall, J. C. Bean, R. People and A. T. Fiory, p. 402, Proc. 1st Int. Symp. on Silicon MBE, J. C. Bean Ed., Vol. PV 85-7, Electrochemical Society Press, Pennington N.J., (1985)

[49] T. P. Pearsall and J. C. Bean, IEEE Elec. Dev. Lett., EDL-7, 308 (1986)

[50] H. Daembkes, H. J. Herzog, H. Jorke, H. Kibbel and E. Kasper, IEEE Trans. on Elec. Dev., ED-33, 633 (1986)

[51] H. Temkin, T. P. Pearsall, J. C. Bean, R. A. Logan and S. Luryi, Appl. Phys. Lett. 48, 963 (1986)

[52] S. Luryi, T. P. Pearsall, H. Temkin and J. C. Bean, IEEE Elec. Dev. Lett. EDL-7, 104 (1986)

[53] T. P. Pearsall, H. Temkin, J. C. Bean and S. Luryi, IEEE Elec. Dev. Lett. EDL-7, 330 (1986)

[54] H. Temkin, N. A. Olsson, T. P. Pearsall and J. C. Bean, Proc. Optical Fiber Communications Conf., Atlanta Ga. (Fed. 1986)

[55] H. Temkin, A. Antreasyan, N. A. Olsson, T. P. Pearsall and J. C. Bean, to be published Appl. Phys. Lett.

Inst. Phys. Conf. Ser. No. 82
Paper presented at ESSDERC 1986, Cambridge 8–11 Sept. 1986

Pure and applied research aspects of dry etching for silicon technology

V G I Deshmukh

Royal Signals and Radar Establishment, St Andrews Road, Great Malvern, Worcestershire, WR14 3PS, UK.

Abstract. Dry etching has become an accredited technique in the fabrication of silicon-based microelectronic circuits where critical dimensions are less than about three microns because the anisotropic nature of the process allows superior pattern fidelity over wet chemical methods. Using examples from the research programme at RSRE, this paper reviews research studies both in practical applications of dry etching and experiments designed to elucidate some of the underlying physics and chemistry of dry etch processes. With regard to exploitation of the dry etch technique, the interplay is highlighted between the etch process, the nature of the material being etched, the lithographic pattern definition stage and final device and circuit requirements. On the pure research side, diagnostic experimentation is also described which is shown to be capable of explaining practical etching artefacts and behaviour.

1. Introduction

As a usable technique, dry etching has become an indispensable part of a semiconductor process engineer's armoury for the production of advanced microelectronic circuits where critical dimensions are less than about 3 microns. As a subject for study, dry etching has become sufficiently mature for a number of detailed treatments to appear on the library shelves, for example those of Chapman (1980) and the texts edited by Einspruch and Brown (1984) and Powell (1984). Despite these facts, interest in dry etching is far from moribund. Something over one thousand papers have been published in the field and, undoubtedly, additional accounts exist that remain undistributed on the grounds of commercial confidentiality. Nor yet is there a sign that the number of annual publications concerning dry etching is on the wane, since the understanding of existing processes is still largely empirical and new materials and techniques are constantly under evaluation. The purpose of this paper is to give some typical illustrations of the problems and achievements associated with the application of dry etching to silicon device fabrication and to describe representative experiments that can yield insight into the underlying science of the dry etch process. With respect to the applications of dry etching, attention is restricted to silicon-based MOS requirements.

As a broad generalisation, research interests in dry etching fall into pure and applied categories. Examples of the former would include the work of Coburn, Winters and co-workers (1977, 1979) who have made comprehensive studies of ion-surface interactions pertinent to dry etching systems under conditions of ultra high vacuum (UHV) and the detailed chemical studies of dry etching mechanisms undertaken by Flamm et al (1984). On the applied research side, a plethora of papers exist that demonstrate the etching of thin films and substrates of interest and the dependence of the dry etch performance on such parameters as gas composition and flow rate, working pressure, power density and so on. At first sight, it is remarkable that so many man-years' work in so many different laboratories worldwide has been necessary to accomplish desired dry etching processes for a rather limited range of materials. Aside from the obvious commercial interests and the dependence of the required etch process on the base technology being addressed, more subtle factors must be in play.

Firstly, the materials may be small in number but they are diverse in morphology, conductivity and composition. Thus we must consider single-crystal and polycrystalline materials such as silicon and polysilicon, amorphous films or glasses (eg silicon dioxide) and polymeric layers, eg photoresists. Material conductivity ranges from insulating or semi-insulating through semiconducting (from intrinsic to degenerate) to full metallic values.

Compositional variations span those peculiar to chemical compounds such as silicides, alloys and films of inconsistent stoichiometry, eg silicon oxynitride, $Si_xO_yN_z$. For a given etching environment, the etch performance has been found to vary according to the morphology, conductivity and composition of the films of interest. Moreover, combinations of materials are nearly always present simultaneously in an etching process. An example would be a defined polymeric material (photoresist) atop a metallic conductor (Al:Si:Cu) deposited on silicon dioxide. The presence of the organic material can cause changes in the anisotropy of the etching of the aluminium alloy through redeposition on the feature sidewalls.

Secondly, the dry etch process cannot be taken in isolation. The fabrication of any integrated circuit can be considered to be a series of interlocking topics in applied science. Those technologies of particular relevance to a pattern transfer stage are film deposition or growth, lithography and etching. The quality of a dry etching process is intimately related to the preceding deposition and lithographic stages. Thus, improved etching performance and relaxed tolerancing of the etch may sometimes be achieved by improvements in the deposition and lithographic stages rather than the etch process itself.

Thirdly, the dry etching process is extremely complex. Apart from the genuine problems of attempting to unravel the detailed physics and chemistry of a typical dry etch utilising a multicomponent gas plasma, other seemingly prosaic, but none the less real difficulties lie in the choice of reactor geometry, pumping fluids, gas purities, etc.

Overall, these features conspire to make the quantification of a dry etch process particularly intransigent both in terms of the quality of performance and the dependence of the performance on the physical parameters that define the etch process. It follows, to borrow a term more often applied to computer software, that the "portability" of a dry etching process is often poor.

Notwithstanding these difficulties, dry etching is a vital tool of the semiconductor fabrication industry and the following sections will outline

TABLE 1: IDEAL DRY ETCHING CHARACTERISTICS

Anisotropy:	High; or as required in a sloped sidewall process.
Selectivity:	High.
Resist Erosion:	Zero; except in planarisation or resist erosion etch cycles.
Damage:	Zero or recoverable, eg by thermal anneal.
Applicability:	Wide; independent of conductivity, morphology and composition of the material.
Uniformity:	High, within wafer, wafer-to-wafer.
Reproducibility:	High, run-to-run.
Throughput:	High.
Equipment:	Simple, easily maintainable, safe.

examples of the achievable performance and attendant problems of present etch systems and of interrogative experiments that can yield insight into the underlying physics and chemistry of the process.

2. Applications of Dry Etching to Silicon Device Fabrication

The basic components of a modern dry etching machine are a vacuum chamber with associated pumping system, an inlet control system for the reactive gases of choice and a radiofrequency supply feeding power, via a matching network, to electrodes internal to the vacuum space. In operation, a plasma is struck in which the ion density lies between 10^{10} and 10^{11} cm^{-3} and the ratio of ionic to neutral densities is $1{:}10^{4}$. Any surface, including the wafer, in contiguity with the plasma is separated from it by a dark space or sheath supported by an electric field, typically several hundred volts/cm, that can accelerate positive ions from the plasma-sheath edge towards the wafer. Etching proceeds principally by the formation of volatile products, either by purely chemical reaction between adsorbed reactive species on the wafer surface or by physico-chemical reactions in which product formation or desorption is stimulated or controlled by ionic bombardment. Purely physical etching, more readily recognised as sputtering, can also occur. The fact that the sheath field dictates that ion bombardment is normal to the wafer surface is consequential in yielding an anisotropic etch process, ie the 'vertical' etch rate is larger than that laterally. An additional chemical effect which can also occur is the redeposition of products onto the surfaces of features being etched. These deposits protect the surfaces except where removed by ion bombardment, that is at the bases of features which are normal to the directed ion motion rather than the sidewalls which are parallel to the ions' paths. Hence, sidewall deposition may also play a role in ensuring an anisotropic quality of a dry etch process.

An ideal dry etching process would have the attributes shown in Table 1. It is difficult to optimise a particular etch process in all respects and a compromise is generally sought in which the most important characteristic is accentuated. Moreover, the quantification of a particular parameter descriptive of the etch performance is dependent on

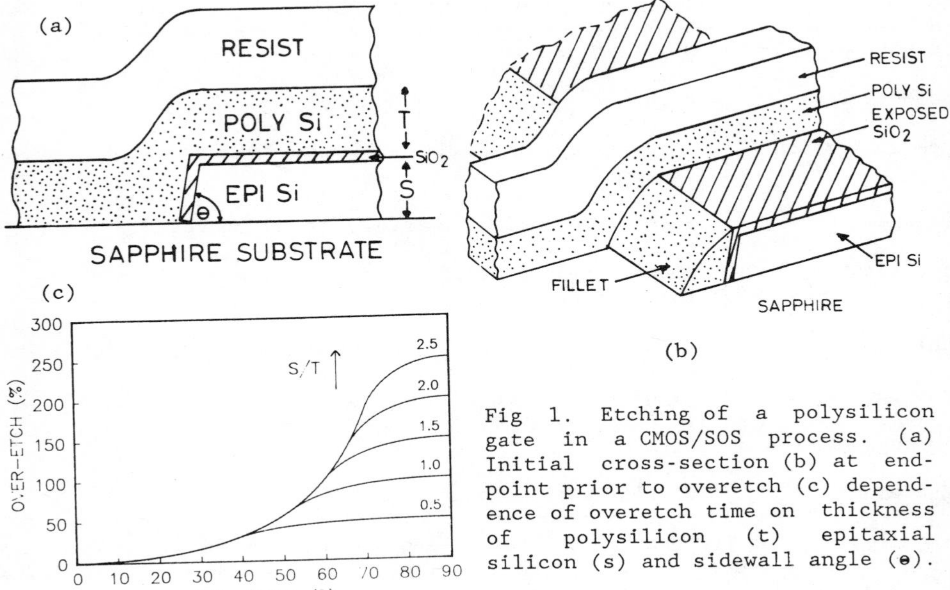

Fig 1. Etching of a polysilicon gate in a CMOS/SOS process. (a) Initial cross-section (b) at end-point prior to overetch (c) dependence of overetch time on thickness of polysilicon (t) epitaxial silicon (s) and sidewall angle (θ).

the application of the said process. For example, consider the trade-off between selectivity and anisotropy in the definition of a polysilicon gate in a CMOS/Silicon-on-Sapphire (SOS) technology. The selectivity of a dry etch process may be defined as the etch rate ratio of film to sublayer or substrate. For a wet chemical process, of course, the selectivity can be infinite but a typical value in a dry process is of order 10-50. As mentioned above, the base fabrication technology will determine the requirements of a dry etch and figure 1a shows a SOS wafer at the polysilicon gate definition step. For a modern SOS process with a transistor gate length ~ 1 µm, clearly the anisotropy of the gate etch, that is the ratio of vertical to lateral etch rates, must be kept high. This is achieved by high energy ion bombardment and will therefore result in diminished selectivity due to the chemical insensitivity of ionic sputtering. However, the combination of an anisotropic etch process and the effective increase in thickness of the polysilicon over the island edges (fig 1a) implies that an overetch period is required to clear the polysilicon fillet (fig 1b) at the base of the silicon island without breaking through the gate oxide which is 10-30 nm thick for a small-geometry process. During overetch, the required selectivity is of order 30:1. The overetch time is a function of the thickness of the epitaxial silicon, the thickness of the polysilicon layer and the angle of the island sidewall (fig 1c). These quantities also strongly impinge on the transistor and circuit properties. Thus, the carrier mobility is a function of epitaxial silicon thickness and a value of ~ 0.3 µm is commensurate with the maintenance of respectable mobility and device gain. Clearly, the resistivity of the doped polysilicon gate and interconnect layer is dependent on thickness and a value ~0.5µm thickness is customary Finally, the island sidewall angle affects device properties since the parasitic sidewall transistor has a threshhold voltage which is a function of the sidewall interface state density and this is in turn dependent on the crystallographic orientation of the silicon and thus the wall angle. Overall, then, the gate dry etch process is defined by factors pertaining to the whole process and the electrical device characteristics rather than isolated estimation of the required values of selectivity and anisotropy.

As far as resist erosion is concerned, consider the extreme case of a sub-half micron gate FET. The best known method for definition of the resist layer at these very small dimensions is via electron beams. By their very nature of being exposed by particle beams, electron-sensitive resists, not unsurprisingly, are easily degraded by dry etching plasmas. Moreover, for any resist, image resolution goes hand-in-hand with the thinness of the resist mask and again this is contrary to the requirements of resist longevity in a dry etching plasma. The problem can be overcome by the use of trilevel resist structures consisting of an upper, thin working resist, a thin 'pattern transfer' layer and a lower, thick planarising layer such as polyimide. The drawn resist pattern is etched into the central pattern transfer layer which then acts as a mask for the definition of the planarising layer in an oxygen plasma, as shown in figure 2 (Brown et al, 1985). Thus dry etching is forming an intimate part of the lithographic process as well as finding use for the definition of the polysilicon film underlying the polyimide mask (fig 2). The penalties associated with the increased processing time for a trilevel resist stage and the serial exposure nature of electron beam lithography are unacceptable in a manufacturing environment at present. Many modern pilot or production MOS processes have a drawn gate dimension ~ 0.8 - 1 μm. At this geometry, optical lithography using direct-step-on-wafer exposure tools and single-level photoresists are usable.

(a)

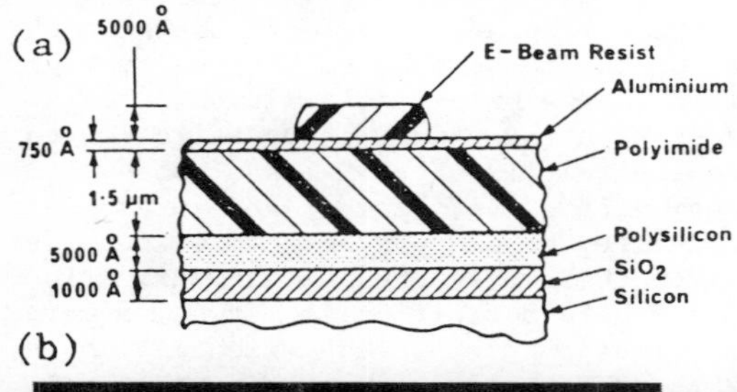

Fig 2. (a) Trilevel resist scheme, (b) Electron beam resolution test showing anisotropic definition of polyimide, (c) anisotropic etching of polysilicon using polyimide mask of (b).

(b)

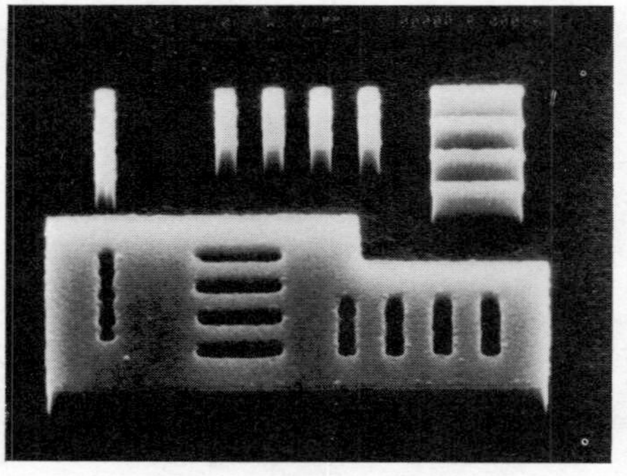

(c)

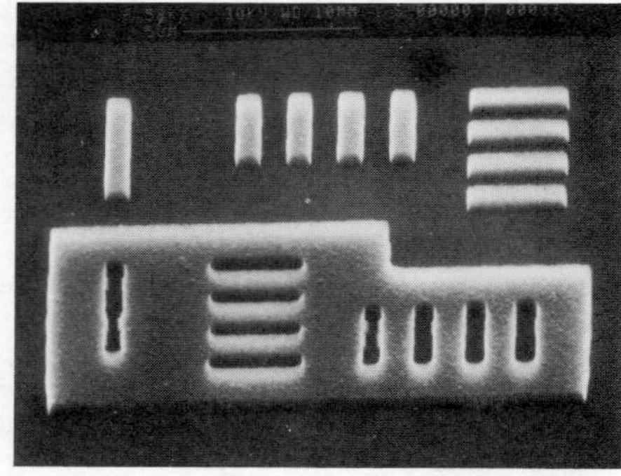

The role of the photoresist during dry etching is not well understood but there is evidence that photoresist erosion products influence the etch process and, if deposited on feature sidewalls, can lead to an anisotropic process as mentioned above (Bruce and Malafsky, 1982). Plasmas, particularly those containing chlorine, may severely damage photoresists so that the etch rate ratio of film to resist may be undesirably low. During polysilicon etching in a chlorine-containing plasma, however, it is possible under certain conditions for a durable capping layer to be formed on the surface of the resist. The layer is very thin and thus not readily analysed but is conjectured to be adsorbed silicon subhalide species $SiCl_x$ forming a complex with the resist. If oxygen is subsequently added to the etch gas mixture, a thin SiO_x layer may be formed on the resist. In these circumstances, the photoresist etch rate is effectively zero (Scott et al, 1986) and, in an anisotropic process, essentially perfect fidelity of polysilicon etching to the resist mask can be obtained. More generally, the ability of the photoresist to withstand an aggressive dry etch process may be improved by an ultraviolet hardening treatment. Following definition of the photoresist, blanket UV exposure causes increased cross linkage thereby stabilising the drawn features during a subsequent bake at a temperature greater than the normal glass transistion temperature.

Without the blanket UV exposure, the post bake at elevated temperature would lead to resist flow. This flow property can be used to advantage, however, in the definition of circuit features with sloping sidewalls such as contact holes in a dielectric layer based on silicon dioxide. This insulation often consists of a lower thermal oxide (grown at the gate oxide stage) overlaid by a pyrolitically deposited oxide with a surface layer doped with phosphorous and/or boron to several weight per cent. The flowed resist/oxide structure is dry etched in a plasma consisting of a fluorine-containing gas, such as CHF_3 and oxygen such that the etch rate ratio of resist and oxide is of order unity. In this way the flowed, sloped resist profile is reproduced in the underlying oxide in a single dry etching stage. The problems associated with the resist flow technique are in the reproducibility of the flow characteristics over a wafer, particularly with minimally-sized contact holes and that surface tension effects can lead to a dependence of the wall profiles on both the size of opening and the local packing density of the contact holes. Dry etching processes can be developed which yield sloped oxide walls without resist flow by using a multiple step process with individual steps tailored to deal with successive layers of the composite film of the glass, deposited and thermal oxides. By suitable choice of gases, power and pressure, almost any desired concavity or convexity of the contact hole may be created to satisfy the requirements of the subsequent metal definition sytem. The trade off, then, for contact hole definition is between a more complex multistage etch cycle or a one-step etch process which may have resolution limitations. Ultimately though, the success of a contact hole etch is determined less by the shape than by the contact resistance. Additional processing such as removal or consumption of any damaged silicon within the contact region may be necessary to ensure that this is achieved.

Turning to the dependence of etch performance on the electrical and morphological properties of film being etched, two clear examples are provided by doped polysilicon and aluminium alloy films. The etch rate of polysilicon depends on the dopant element, doping density and degree of activation of the dopant. More specifically, the etch rate is higher for n^+ polycrystalline silicon, intermediate for intrinsic material and lowest for p^+ polysilicon, fig 3. The actual enhancements are dependent on the etch system used and tend to be appreciably greater in the higher pressure etch processes (Baldi and Beardo, 1985). This has been explained by the fact that the n-type doping raises the Fermi level and reduces the energy barrier for electron transfer between the solid and an adsorbed halogen atom on the film's surface (Flamm et al, 1984). This in turn leads to an increasing possibility of etchant species penetrating into the film and enhancing the etch rate. Of more concern to the process engineer, however, is the fact that the etch rate enhancement is not directional so that undercutting of the n^+ polysilicon beneath the mask region can occur.

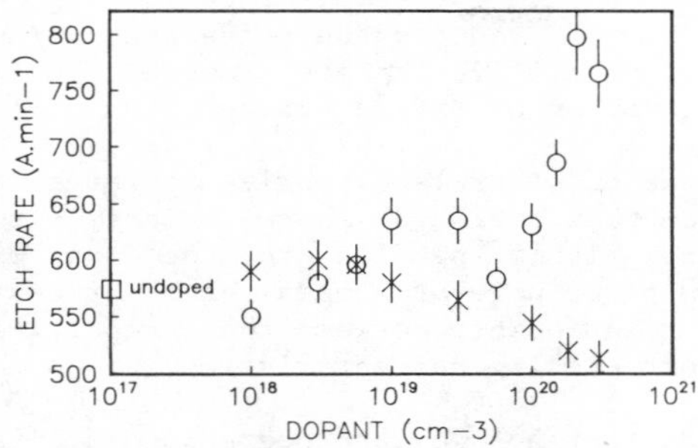

Fig 3. Polysilicon etch rate dependence on implanted dose of phosphorus (o) and boron (*) after activation anneal.

With regard to aluminium metallisation, an alloy containing 1-2% Si and 0.5-4% Cu is often used and may be etched in a plasma containing reactive chlorine species. The copper chlorides are not very volatile and physical sputtering certainly aids in the removal of copper during a dry etch. Etch performance can be strongly dependent on the metal deposition method, which, for a ternary alloy, is most easily performed in a sputterer. This dependence of etch performance on deposition follows since the stoichiometry of the alloy can vary through the film thickness depending on deposition conditions such as substrate temperature, rate of deposition and the applied bias. In a worst case, the copper can migrate through the film to the alloy/subsurface interface forming a copper-rich layer which may be very difficult to remove. With non-bias sputtered material, the lack of surface atomic movement during the sputter deposition can also lead to microcracks in the metal film at the base of step features on the wafer surface. Apart from the deleterious effect of microcracks on track resistivity, we have found in our experience (Crabb et al, 1986) that material fillets at the base of surface steps can be difficult to clear completely in non-bias sputtered material without unacceptable overetch times. It seems natural to relate the problem of fillet removal with the presence of microcracks though this is difficult to justify formally. Possibly, copper migrates to the microcrack to give a locally-enhanced copper composition of the alloy or alternatively, the standard photoresist development solutions may locally infiltrate and attack the microcracked region. At any event, bias-sputtered aluminium alloy can be successfully etched as shown in fig 4 and exhibits better step coverage and lower electrical step resistance than the non-bias-sputtered variety and is

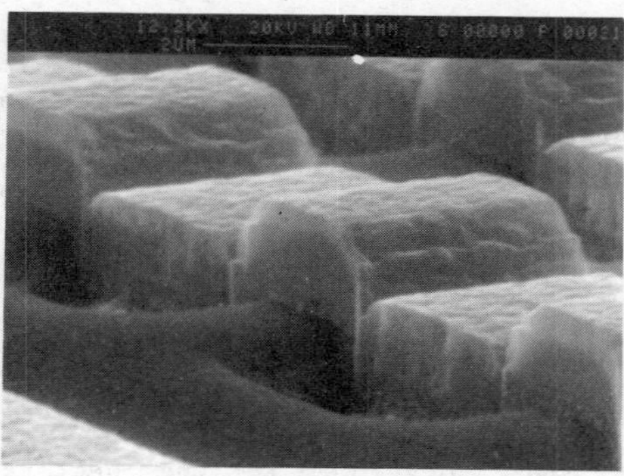

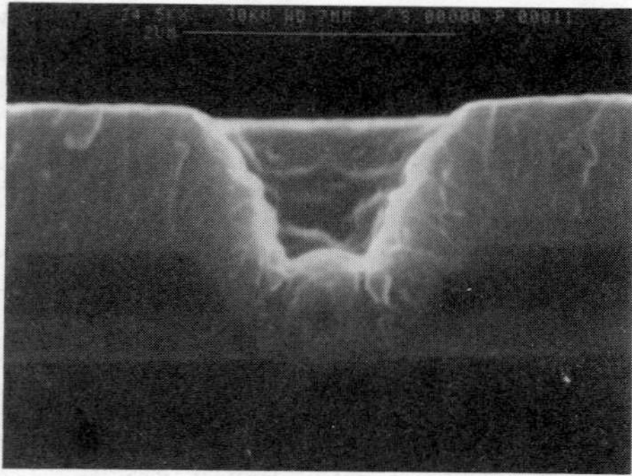

Fig 4. Anisotropic etching of Al:1.5% Si:4% Cu lines over SiO_2 and a cross section through a contact hole etched in SiO_2 with the same metallisation.

therefore preferred. The gases used to etch the sample shown in fig 4 were a BCl_3, Cl_2 and CF_4 mixture. The BCl_3 reduces the oxide present on the surface of the aluminium film thereby facilitating its removal. BCl_3 also acts as a scavenger for any water vapour or oxygen present.

It is quite likely that the experience of other laboratories in etching aluminium alloys may be different from that outlined above, since both deposition and etching are obviously highly specific to the systems employed. However, the description given here of a metal etch process furnishes a good example of the close relationship between the properties of the deposited film and its subsequent etching characteristics.

So far, the assumption has been made that no device degradion occurs during a dry etching process. However, a number of papers exist which show that ion bombardment can cause damage in semiconductor devices both through lattice disorder or by the creation of traps in silicon dioxide (Ephrath and DiMaria 1981, Fonash 1985). Experiments at RSRE (Deshmukh, 1982) have shown that when oxidised silicon wafers are exposed to bombardment by positive ions at 1 keV, then a substantial increase in measured oxide charge and interface state density occurs which scales with ion fluence. The interface state density shows a maximum at midgap. Simple MOSFETs incorporating an ion-bombarded gate oxide demonstrated impaired turn-on behaviour, degradation in the transconductance-gate voltage curve, and a threshold voltage that scaled with ion dose. Fortunately, in a real device process, the active region of the transistor is protected by a gate conductor which guards against trap or mobile ion generation. Furthermore, a forming-gas anneal is often used as a metallisation sinter and this seems to be particularly efficacious in healing any damage created in devices. The possibility of contamination is also present in a dry etching process since etching or sputtering of all surfaces exposed to the plasma can occur. Thus, for example, heavy metals sputtered from chamber walls may migrate to the wafer and subsequently lead to device deterioration. Such contamination may be minimised by astute reactor design. Thus, materials with low sputtering yields are preferred within the etcher and sheath fields are arranged to be small at exposed metal surfaces to reduce the possibility of sputtering still further. Perhaps the most obvious manifestation that damage due to dry etching can be <u>avoided</u>, is simply the countless number of integrated circuits produced worldwide each year in which all pattern transfer stages are dry.

This section has attempted to show that the development of useful dry etching processes is based on the requirements of the device and circuit properties as well as the nature of the lithographic process and the film deposition or growth. The successful implementation of such dry processing is based on a sound appreciation of this 'product'/process interface. The issues of reproducibility, uniformity and avoidance of particulate contamination are vital to whole silicon processes but outside the scope of this paper. In the following paragraphs, an examination of some underlying physics and chemistry of particular dry etching process is given.

3. <u>Examples of Fundamental Research in Dry Etching</u>

Whilst dry etch processes can be successfully developed for the pattern definition stages of contemporary silicon MOS processes, a detailed understanding of the physics and chemistry of the plasma/solid interaction for the gas combinations and the thin films of interest remains undetermined. As mentioned above, incisive studies of ion-surface interaction in UHV or of chemical reaction kinetics have been made but the link remains to be forged between ideal and real dry etching systems and much work remains to be done. Not least of such work is the confirmation that experiments conducted under ideal conditions are of actual relevance to real systems using moderate vacua, complex gas mixtures and samples having variable quality in their physical and chemical properties. Zarowin (1984) has chosen to model dry etching by assuming a plasma of fixed geometry and relating observable etch characteristics such as selectivity or anisotropy to ion transport in the sheath fields. At RSRE, we have chosen to investigate experimentally real etch systems though of necessity the gas/substrate combinations are simplified and the etch systems themselves fall short of true production etch reactors.

As example, the first system is a reactive ion beam etching (RIBE) station in which a well characterised ion beam (0-2 kV) is created in a Kaufman source mounted on a work chamber evacuated by a trapped diffusion and mechanical pumps (Cox and Deshmukh, 1985a). The chosen gas inputs were the fluorocarbons CF_4, C_2F_6, C_3F_8 and samples were of silicon or silicon dioxide. The chosen diagnostic techniques were those of optical emission spectroscopy and quadrupole mass spectrometry which yield information on the state of the bombarded surface and the products of etch reactions.

Several reasons obtained for the choosing of this physical system for investigation. RIBE has an intrinsic interest as an etch method because of its high anisotropy and can also be considered to bridge the gap between UHV and production-oriented dry etching environments. Indeed, pioneering studies of the interaction between flurocarbon ion beams and Si in UHV already existed in the literature (Coburn et al, 1977). Additionally, etch rate data for fluorocarbon RIBE of Si existed in the literature (Heath, 1982) which showed clearly that a threshold in ion energy existed, above which significant enhancements of etch rate existed. The value of the said threshold energy was dependant on whether a cryopump or untrapped diffusion pumped system was used. Thus, this system provided an attractive testbed to determine whether spectroscopic diagnostic techniques could be used to elucidate practical etch behaviour. Experiments showed (Cox and Deshmukh, 1985a) that fragmentation of the bombarding $C_xF_y^+$ ion occurred on impact. The carbon was then adsorbed on the surface leading to carbonaceous passivation of the silicon wafer. Thus, before rapid silicon etching can proceed, the carbon must be removed by combination with F or by physical sputtering and both of these removal paths become more favourable at higher ion energy. In particular, rapid silicon etching occurs above a theshold corresponding to the ion energy at which the surface is only partially passivated by carbon. This behaviour precisely mirrors the etch rate characteristics described above. Taking carbon removal as the principal factor in determining the onset of silicon etching also leads to the anticipation that the associated threshold energy will decrease as the F/C ratio increases in the parent etch gas, and this was indeed observed (Cox and Deshmukh, 1985b). Furthermore, the dependence of etch rate/ion energy curves on the pumping system suggested that the presence of water vapour or oxygen could play an important role in the etch process. Accordingly, oxygen was bled in a controlled way into the etch chamber and the threshold voltage for silicon etching was found to decrease as the partial pressure of oxygen was increased. This could be explained by the additional carbon removal paths as volatile CO or CO_2 provided by the oxygen addition. Additionally, a silicon dioxide sample, having, in effect an inbuilt oxygen supply, could be considered to be equivalent to a silicon wafer etched in an infinite oxygen limit. Our experiments confirmed that the threshold energy for the removal of passivating carbon from SiO_2 was close to that of Si with an appreciable background presence of oxygen gas in the work chamber. This allows the forecast that fluorocarbon RIBE of SiO_2 could exhibit a selectivity over Si only if adequate pumping systems were to be employed to ensure low background oxygen and water vapour. Taken overall, these results show rather convincing evidence of the usefulness of diagnostic techniques in explaining practical artefacts observed in dry etching.

However RIBE is not a standard dry etching technique and the question arises whether diagnostic techniques may be used to advantage with the inherently more complex plasma-based etch systems. To test this, experiments have been carried out with a parallel-plate reactor in which the applied rf power may be split between the electrodes from 10-90% (Hope et al, 1986). The chosen etch system was the simple one of oxygen plasma or reactive ion etching of an organic polymer masked by a non-erodible metal film. Though a simple system, it has important practical application in a trilevel resist scheme. Conceptually, this dry etching process can be divided into chemical and physical components. In the described experiments (Hope et al, 1986), the total power to the reactor was maintained constant but power splitting to the electrodes was varied to alter the dc bias at the wafer-bearing electrode. The plasma was characterised by the electron temperature, as determined independently by Langmuir probe and optical emission spectroscopic measurements, though only the former provides an absolute value. The electron temperature and electron energy distribution function are, of course, important indicators of the chemical composition in the plasma which is largely determined by fragmentation and excitation caused by energetic electron collisions with the gaseous species. For a variation in dc bias on the wafer-bearing electrode between 0 and 250 V, the measured electron temperature was sensibly constant, so that there is some justification in assuming that the plasma composition is constant and therefore that changes in anisotropy of the etch are a result of changing dc bias alone. That is, the plasma is identified simply as a convenient source of chemically reactive species and an observed etching property, namely the anisotropy, occurs only through ion bombardment as controlled by the sheath electric fields which are related to the dc bias on the driven electrode. The neutral species are unaffected by sheath fields and thus the isotropic etch component is unaffected by dc bias. The vertical etch rate may then be written as the sum of the isotropic component due to neutral/solid interactions (and therefore constant for this experiment) plus a bias-dependent, ion-enhanced etch rate. The latter is in turn the product of the ion flux and the depth of material removed per ion which may be determined respectively for different ion energies from the ion current to the Langmuir probe at different biases and the etch rate versus energy characteristic for the etching of the polymer film in an oxygen (O_2^+) ion beam. The deduced anisotropy can then be compared to the measured values determined from measurements of cleaved samples in a scanning electron microscope. In view of the fact that charge exchange in the sheath and ion-stimulated reactions have been ignored, the agreement obtained is surprisingly good (fig 5). In spite of this agreement, some cautionary words are needed. Firstly, Langmuir probes are invasive to the plasma and notoriously prone to misinterpretation due to perturbation of the plasma, contamination and/or etching by the plasma and rf pickup at the probe. The net result can lead to seriously erroneous estimates of the electron temperature. Fortunately, for a balanced rf system such as available for our experiments, rf interference can be eliminated and dc probe theory applied. Alternatively the probe may be radiofrequency driven by picking off part of the rf supply to the chamber electrodes (Braithwaite et al, 1986). Both methods give closely similar results for electron temperature in our apparatus and values of $T_e \sim 4$eV are typical.

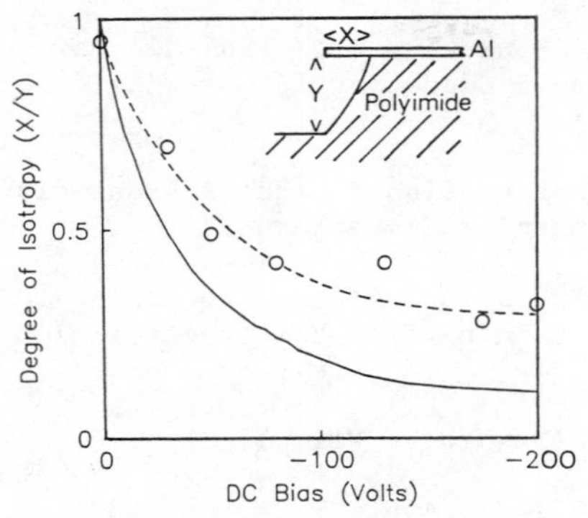

Fig 5. Model (-) and measured (o) anisotropy for the etching of polyimide as a function of DC bias.

The electron energy distribution may be obtained from the second derivative of the probe I-V characteristic and results for argon and helium plasmas show that the distribution is closely Maxwellian for our system at pressures around 20 millitorr. At lower pressures, a group of electrons with substantially higher energies (>20eV) can be observed. Currently, these experiments are being extended to determine the relationship between etch anisotropy and ion/neutral composition of the oxygen plasma at constant dc bias and to see whether Langmuir probe measurements can be used to aid the transfer of processes between etch reactors of different marque.

4. Conclusions

Dry etching equipment and processes are now available which allow the definition of features useful for modern MOS processes. These have evolved via an appreciation of how a dry etching stage is influenced by preceding process steps and how it itself affects final device performance. The detailed scientific understanding presently lags the capability demonstrated in many fabrication lines. However, we hold that the underlying science is necessary if the technique is to be advanced to fulfil the more stringent requirements imposed by ultra small devices and to become more widespread. Diagnostic tools on etch machines will certainly be necessary if process portability or upscaling is required since the plasma properties must be transferred or scaled as well as such parameters as power density, gas residence time, etc. Moreover, the move to completely automated device fabrication lines implies that real-time monitoring of the plasma environment would be desirable as well as the more commonly monitored variables such as rf power, dc bias, etc. Whilst, as noted above, Langmuir probes can provide direct information on the plasma parameters, the inability of probes to withstand common etch gas mixtures implies that a non-invasive technique such as optical emission spectroscopy is rather more likely to be considered compatible with clean room operation.

5. Acknowledgements

It is a pleasure to acknowledge the contributions of colleagues within the Signal Processing Group at RSRE, particularly T. I. Cox (who also provided a critical reading of the manuscript), J. G. Crabb, D. A. O. Hope, S. J. Till and A. G. Brown. The support and encouragement of A. L. Mears and G. R. Jones is appreciated. Finally, the author is indebted to J. E. Allen, N. StJ. Braithwaite and N. M. P. Benjamin of the Engineering Department, University of Oxford, for considerable help with Langmuir probe measurements.

6. References

Brown A.G., Mortimer S.H., Till S.J. and Deshmukh V.G.I. 1985 Microcircuit Engineering 85 ed. K.D. Van der Mast and S. Radelaar (Amsterdam: North Holland) pp 443-450.

Braithwaite N.StJ. 1986 IEE Vacation School on Plasma and Plasma-related Etching of Semiconductor Materials (London: IEE) Lecture 12

Bruce R.H. and Malafsky 1982 Proc. Third Symp. Plasma Processing ed. J. Dieleman, R.G. Frieser and G.S. Mathad (Pennington: Electrochemical Soc) pp 336-347.

Chapman B. 1980 Glow Discharge Processes (New York: Wiley).

Coburn J.W., Winters H.F. and Chuang T.J. 1977 J. Appl. Phys. 48 3532.

Coburn J.W. and Winters H.F. 1979 J. Vac. Sci. Technol. 16 391.

Cox T.I. and Deshmukh V.G.I. 1985a Appl. Phys. Lett. 47 378.

Cox T.I. and Deshmukh V.G.I. 1985b RSRE 1985 Research Review (RSRE MOD(PE): Malvern) pp 146-150.

Crabb J.G., Cox T.I. and Deshmukh V.G.I., to be published.

Deshmukh V.G.I. 1982 IEE Colloq. 'Dry Etching Related to Silicon' (London: IEE) p 4/1-2.

Einspruch N.G. and Brown D.M. (eds) 1984 VLSI Electronics Microstructure Science 8 Plasma Processing for VLSI (London: Academic).

Ephrath L.M. and DiMaria D.J. 1981 Solid State Technol. 24 183.

Flamm D.L., Donnelly V.M. and Ibbotson D.E. 1984 VLSI Electronics Microstructure Science 8 Plasma Processing for VLSI ed N.G. Einspruch and D.M. Brown (London: Academic) pp 189-251.

Fonash S.J. 1985 Solid State Technol. 28 201.

Heath B.A. 1982 Proc. Third Symp. Plasma Processing ed. J. Dieleman, R.G. Frieser and G.S. Mathad (Pennington: Electrochemical Soc) pp 421-435.

Hope D.A.O., Cox T.I. and Deshmukh V.G.I. 1986 Vacuum (in press).

Powell R.A. (ed) 1984 Materials Processing - Theory and Practices 4 Dry Etching for Microelectronics (Amsterdam: North Holland).

Scott S.C., Hope D.A.O., Cox T.I. and Deshmukh V.G.I. 1985 RSRE Memorandum, No 3894 in the press (MOD(PE): Malvern).

Zarowin C.B. 1984 Materials Research Soc. Symp. Proc. 38 Plasma Synthesis and Etching of Electronic Materials ed. R.P.H. Chang and B. Abeles (Pittsburgh: Materials Research Soc) pp 127-142.

Inst. Phys. Conf. Ser. No. 82
Paper presented at ESSDERC 1986, Cambridge 8–11 Sept. 1986

III–V semiconductors on Si substrates

Russ Fischer† and H. Morkoc

University of Illinois at Urbana-Champaign, Coordinated Science Laboratory, 1101 W. Springfield Avenue, Urbana, IL 61801 USA

Abstract. Recent advances in lattice mismatched epitaxial technology has allowed the demonstration of GaAs based devices on Si substrates with performance comparable to their counterparts on GaAs substrates. One micron gate GaAs MESFETs on Si with transconductances of g_m = 180 mS/mm, and current gain cutoff frequencies of f_T = 13 GHz have been obtained. These values are comparable to similar geometry state-of-the-art GaAs MESFETs on GaAs substrates. Heterojunction bipolar transistors (HBTs) with f_T = 30 GHz and f_{max} = 11 GHz for 4 x 20 μm^2 emitters have been obtained on Si, and compare favorably with values for HBTs on GaAs.

1. Introduction

The potential benefit of GaAs growth on Si substrates has precipitated a recent surge of activity in this area (Fischer et al, 1986b, 1986a, 1986c, Aksun et al). The use of Si as a substrate for GaAs epitaxy would allow larger diameter, better thermal conductivity and better mechanical strength. It would also be possible to monolithically integrate (Fischer et al 1985b) GaAs devices and Si devices taking advantage of the speed of GaAs and the mature processing technology for Si. Optical signals could be generated in such a monolithic system, making optical off-chip communication possible.

There are two main difficulties associated with growth of GaAs on Si. The first is that of lattice mismatch — GaAs has a 4% larger lattice constant than that of Si. This large mismatch will unavoidably cause misfit dislocations. Approximately, a misfit dislocation density of 10^{12} cm^{-2} is required at the substrate-epi interface to accommodate the mismatch. Inevitably, a certain fraction of these will become sources for threading dislocations. Techniques which confine the dislocation lines to the bottom of the epitaxial structure are required.

The second difficulty is the problem of antiphase disorder. The zincblende lattice is composed of two interpenetrating FCC sublattices, one for the cation and one for the anion. Since Si is a nonpolar semiconductor (i.e., there is no distinction between sublattices), uniform ordering of the polar semiconductor sublattices is not ensured. Upon crossing an antiphase boundary, the sublattices are interchanged. Again, growth techniques to suppress this type of defect are needed. In this paper, several growth techniques whereby these problems can be avoided will be outlined. Device results will also be presented to substantiate their effectiveness.

† Present address: AT&T Bell Laboratories, 600 Mountain Avenue, Murray Hill, NJ 07974, USA.

2. Materials investigation

Although the exact electrical nature of antiphase disorder is not well understood, it would be preferable to eliminate it. At an antiphase boundary, Ga-Ga and/or As-As bonds occur. The result is that locally there must be an excess positive or negative charge. The excess charge would be a scattering center which would lower mobility. It is unclear whether deep levels would result.

Fortunately, the technique of depositing prelayers is effective in suppressing antiphase disorder. Prelayer deposition is where a (100) oriented substrate is exposed to either a Ga beam or an As beam immediately prior to growth. By ensuring uniform coverage of either cation or anion on an atomically flat surface, antiphase domains will not form. This is because in the (100) direction, planes alternate between Ga and As.

In the presence of single atomic layer (or odd numbers of atomic layers) steps, the situation is modified. An antiphase boundary will result at such a step with uniform prelayer coverage. These steps are known to exist in nominally (100) oriented Si (Kaplan 1980). Several techniques have been used to detect antiphase disorder. These include plan-view transmission electron microscopy (TEM), X-ray scattering (Neumann et al 1986), and anisotropic chemical etching (Fischer et al 1985a). We have not seen evidence for antiphase disorder by any of these techniques.

Most chemical etches for GaAs attack the {111}B face more rapidly than the {111}A. The {111}A face therefore tends to determine the sidewall shape of a mesa. Upon crossing an antiphase boundary, the {111}A and {111}B faces become interchanged, and antiphase disorder can easily be detected by noting the sidewall shapes. Shown in Figure 1 are scanning electron micrographs of mesas etched in GaAs on Si using an As prelayer (a) and a Ga prelayer (c). Shown in Figure 1(b) is a sample where antiphase disorder was intentionally generated to demonstrate that this technique can reveal antiphase disorder.

The etching technique will probably reveal antiphase disorder if the domain sizes are larger than about 0.2 μm. To detect antiphase domains of smaller size, X-ray scattering was used. The details of this technique are published elsewhere (Neumann et al 1986). Since no antiphase disorder was detected by any technique even in samples grown on nominal (100) oriented Si, there must be some mechanism whereby antiphase boundaries "heal" or are overgrown.

In order to reduce dislocation densities, several techniques for confining them to the lower part of the epitaxial structure were developed and employed. The first of these techniques is the use of tilted substrate orientations. High resolution transmission electron microscopy (TEM) images (Otsuka et al 1986) have shown that when a tilted orientation is used, dislocations with their Burgers vectors parallel to the substrate-epi interface are preferentially generated. In contrast, when non tilted (100) orientations are used there are more dislocations with Burgers vectors inclined with respect to the interface. Since it is only the projection of the Burgers vector onto the interface that accommodates mismatch, dislocation with inclined Burgers vectors are less efficient at taking up mismatch. It is not only that fewer dislocations with Burgers vectors parallel to the interface are required, but also that this type of dislocation does not act as a source for threading dislocations.

The role of tilted substrates is to generate the proper type of dislocations such that they do not propagate up through the epitaxial overlayers. If a tilt towards the <011> azimuth is used, steps only occur in one direction. In contrast, if a tilt in a mixed direction, toward the <001> azimuth, is used two dimensional steps occur. In a substrate with two dimensional steps, dislocations are controlled in both directions so that low dislocation densities result.

Some have suggested that the formation of two dimensional steps in tilting the substrate toward a (001) azimuth will modify the surface such that single

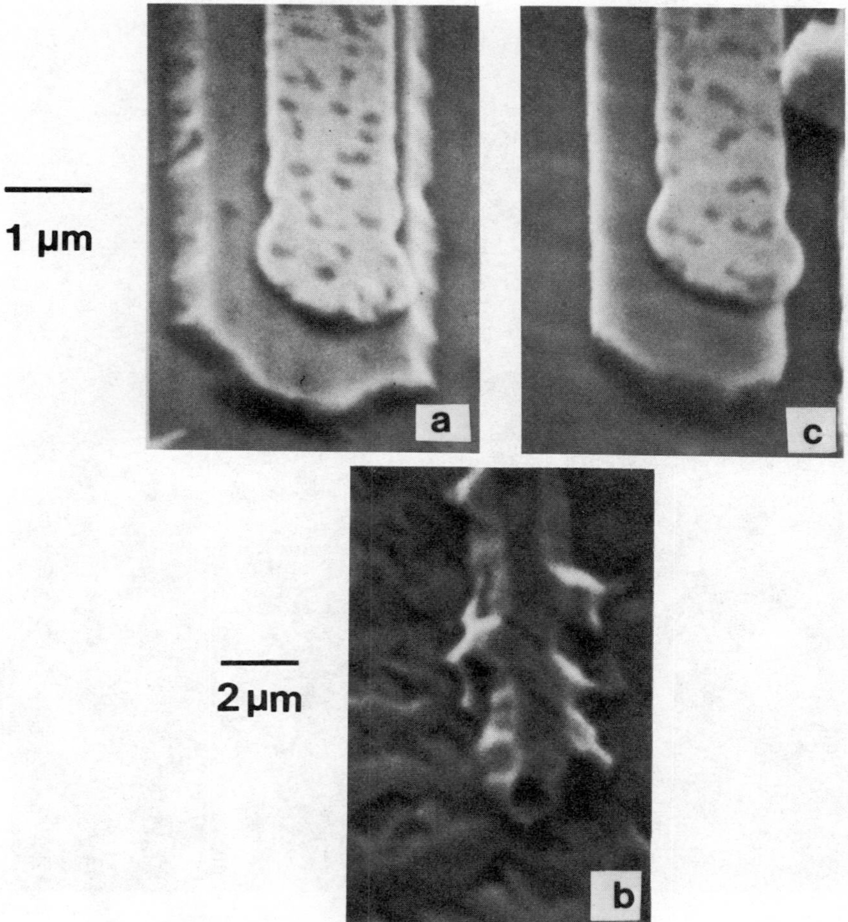

Fig. 1. Sidewall shapes of GaAs epitaxial layers on Si (100), with an As prelayer (a), antiphase domains (b) and a Ga prelayer (c).

atomic layer steps will form (Aspnes). In this case, antiphase disorder would most likely result. We have not found evidence of antiphase disorder in samples grown on substrates with this orientation. Either there is some mechanism where the antiphase domains "heal" or are overgrown, or if there is some inaccuracy in direction of tilt, double steps may still predominate.

It is inevitable that at least a small fraction of the dislocations at the interface will be sources for threading dislocations. Since such large misfit dislocation densities are required at the interface, these will result in an unacceptably large dislocation density in the overlayers. We have therefore used GaAs/InGaAs pseudomorphic superlattices as dislocation barriers. The use of these barriers was predicted in principle by Matthews and Blakeslee (1975). As can be seen in Figure 2, very low densities of misfit dislocations at the top of $\sim 2\ \mu m$ thick films have been obtained using these techniques. From plan view TEM, dislocation free surfaces have been observed, and based on the area of observation the dislocation density must be less than $2 \times 10^5\ cm^{-2}$. We feel that much lower densities have been obtained.

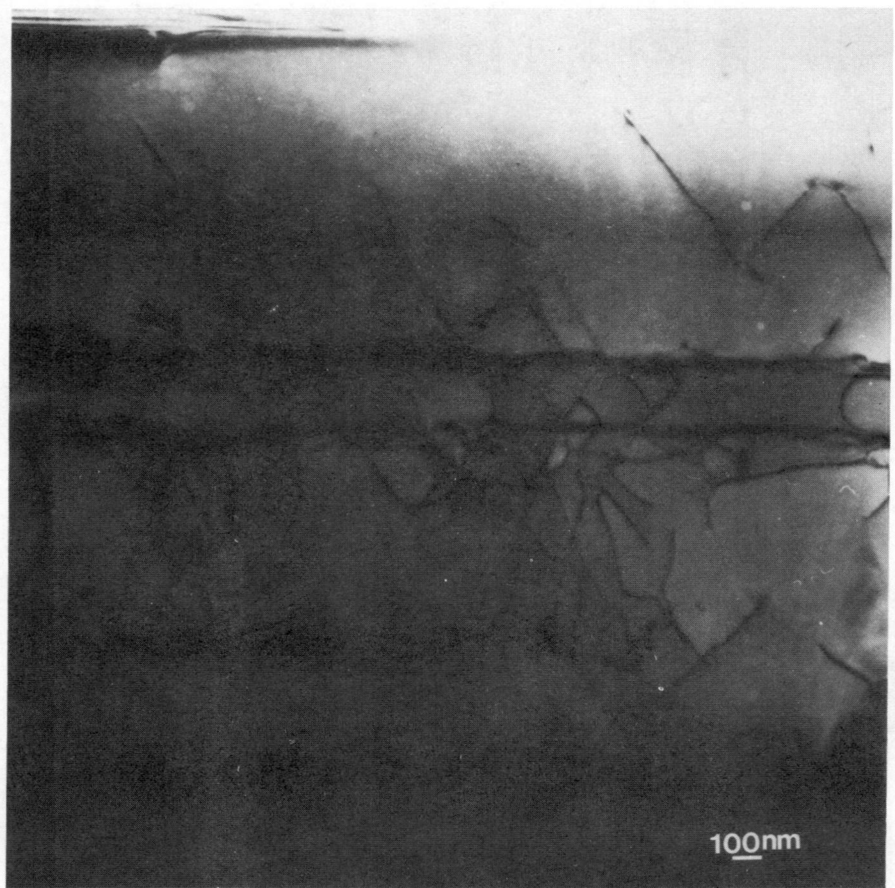

Fig. 2. TEM image of a GaAs on Si epitaxial layer showing the very low misfit dislocation densities.

3. Device Characterization

Many authors have demonstrated high quality GaAs based devices on Si substrates (Fischer et al, 1986b, 1986a, 1986c, Aksun et al). For GaAs metal semiconductor field effect transistors (MESFETs) on Si, performance comparable to state-of-the-art devices on GaAs substrates has been obtained (Fischer et al 1986a). Transconductances of 180 mS/mm for 1.2 μm gate devices on Si have been demonstrated.

It is not only the dc performance that is comparable, but also the performance at microwave frequencies. Figure 3 shows short circuit current gain and maximum available gain as a function of frequency for 1.2 μm gate GaAs MESFETs on Si. There is a current-gain cutoff frequency of $f_T = 13$ GHz and a maximum oscillation frequency of $f_{max} = 30$ GHz. These values compare very well with what has been obtained in comparable geometry transistors on GaAs using both epitaxial and direct implant technology. Equivalent circuit modeling has shown that there is essentially no difference between the devices on GaAs or

Si. All element values are almost identical between GaAs on Si and GaAs on GaAs.

Modulation doped field effect transistors (MODFETs) on Si have also been demonstrated (Fischer et al 1986c). In GaAs/AlGaAs MODFETs on Si, transconductances for 1 μm gate devices of 170 mS/mm and 275 mS/mm at 300 K and 77 K, respectively, have been reported. As in the case of MESFETs on Si, the microwave performance of the MODFETs was nearly identical to their counterparts on GaAs. Current-gain cutoff frequencies of f_T = 15 GHz and maximum oscillation frequencies of f_{max} = 25 GHz were obtained.

Quarter micron gate GaAs MESFETs on Si were also fabricated (Aksun et. al.). Extrapolating H_{21} to 0 dB at 6 dB/octave gave a current gain cutoff frequency of f_T = 55GHz, which is comparable to the best MESFETs on GaAs of similar geometry. This demonstration is significant as it shows that very high frequency performance can be achieved in this system, and that it is possible to fabricate submicron features in GaAs grown on Si. The noise figure at 18 GHz was 2.8 dB with an associated gain of 7.9 dB which is about 1 dB higher than a comparable structure on GaAs. We expect this figure to be reduced with the use of a strained layer superlattice to reduce dislocation densities.

In minority carrier devices, very encouraging results have also been demonstrated. Heterojunction bipolar transistors (HBTs) on Si have been reported with good performance at dc and excellent performance at microwave

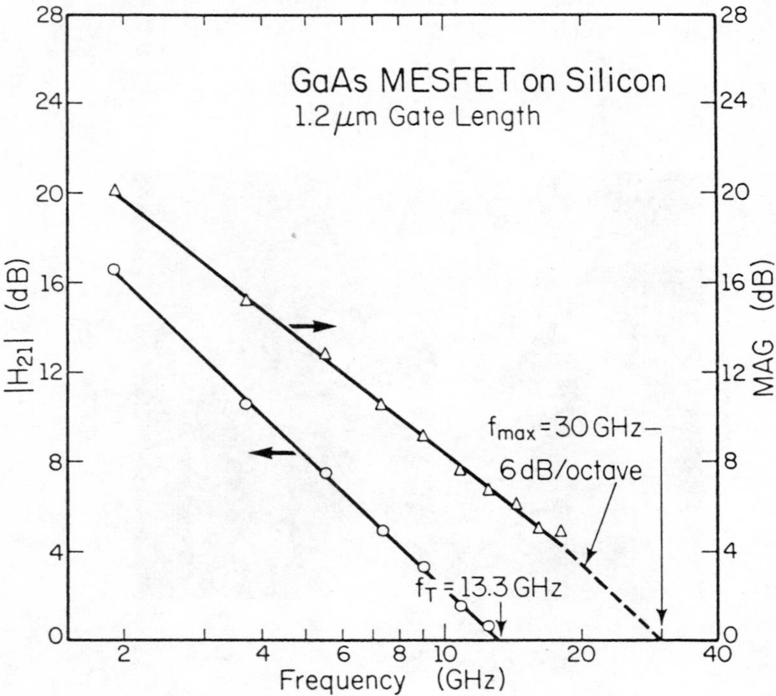

Fig. 3. Short circuit current gain and maximum available gain plotted versus frequency for a 1.2 μm gate GaAs MESFET grown on Si.

frequencies. Shown in Figure 4 is an output current-voltage characteristic of GaAs/AlGaAs HBT grown on Si. As can be seen, a common emitter current gain of $\beta = 13$ is obtained in this device with a 0.2 μm thick base doped to $p = 5 \times 10^{18}$ cm^{-3}. Reduction of defect-assisted space charge recombination and surface leakage component of the base current can lead to larger current gains. The offset voltage in these devices is about 0.2 V, which is expected from the difference in turn-on voltage between the heterojunction emitter and the homojunction collector.

Although the current gains in the HBTs grown on Si are high enough for some applications, they are not as high as would be expected from devices with identical structures grown on GaAs. To investigate the mechanism behind this lower gain, we fabricated a series of HBTs with varying base thickness. As mentioned earlier, defects in the material can lower the minority carrier lifetime resulting in a lower base transport factor. If this were the case, then the gain should increase substantially by reducing the base thickness. However, the gain does not improve upon reducing base thickness as gains of $\beta = 13$, 12, and 13 were obtained in structures with $W_B = 0.2$, 0.15, and 0.1 μm, respectively. This indicates that the gain in these devices is not limited by recombination in the neutral base region. Further studies are in progress to determine the mechanisms which limit the current gain.

As can be seen in Figure 5, maximum current densities of 160 kA/cm^2 were obtained in the HBTs on Si without device failure. Upon operating these devices at current densities above 100 kA/cm^2 for several hours, the current gain did somewhat degrade. However, upon operation for several hours at current densities below about 30 kA/cm^2, no device degradation was observed.

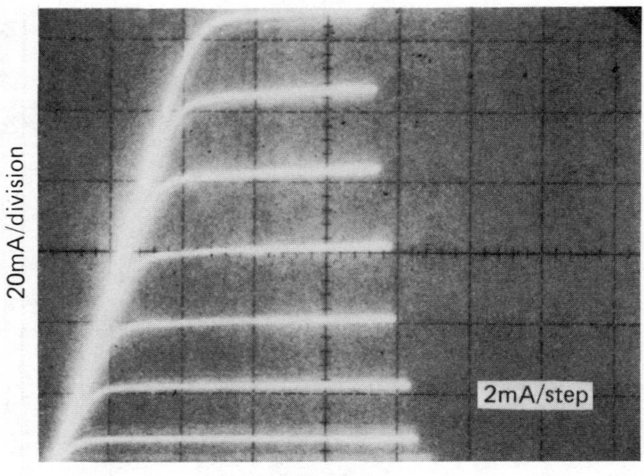

Fig. 4. Output I-V characteristics of an HBT grown on Si.

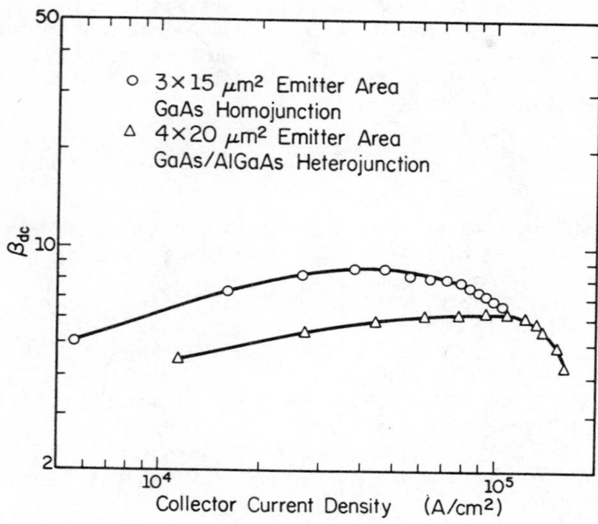

Fig. 5. Current gain versus collector current density for heterojunction and homojunction bipolar transistors grown on Si.

From Figure 5, values for the emitter junction ideality were extracted by taking the slope of the β vs. I_C curve. Ideality factors in the range of n = 1.5 to n = 2 were typical, demonstrating that there is a large space-charge recombination component of base current.

In order to test the high speed performance of these devices, microwave S-parameter measurements were made in the frequency range from 2-18 GHz. For devices with relatively large geometries($4 \times 20\ \mu m^2$) f_{max} values of 11 GHz were obtained with f_T at about 30 GHz at a current density of 15 kA/cm^2. With smaller geometries and optimized structures these figures are expected to improve significantly.

4. Conclusions

In conclusion, growth techniques for GaAs on Si have been developed. It was shown that these techniques are very effective in reducing dislocation densities and eliminating antiphase domains.

Application of these techniques to GaAs devices has resulted in the achievement of state-of-the-art performance. GaAs MESFETs (1.2 μm gate) on Si with transconductances of $g_m = 180$ mS/mm, current gain cutoff frequencies of $f_T = 13$ GHz and maximum oscillation frequencies of $f_{max} = 30$ GHz have been obtained. In HBTs, common emitter current gains of $\beta = 13$ were achieved. Values for f_T of 30 GHz and for f_{max} of 11 GHz were also obtained for a $4 \times 20\ \mu m^2$ emitter geometry. These results are very encouraging and show the potential for GaAs growth on Si.

Acknowledgements

This work was funded by the Air Force Office of Scientific Research of the USA. The authors would like to thank Prof. N. Otsuka and Chin Choi of Purdue University for TEM work, Prof. H. Zabel and D.A. Neumann of the University of Illinois for X-ray scattering work, and M. Todd for manuscript preparation. Technical assistance of W. Kopp, J.S. Gedymin, J. Klem, T. Henderson, and C.K. Peng is also gratefully acknowledged.

References

Aksun M I, Morkoç H, Longerbone M, Erickson L P, Chao P C, Duh K H G, Smith P M and Lester L F Appl. Phys. Lett pending.

Aspnes D E private communication

Fischer R, Chand N, Kopp W, Morkoç H, Erickson L P and Youngman R (1985a) Appl. Phys. Lett. 47 397

Fischer R, Henderson T, Klem J, Kopp W, Peng C K, Morkoç H, Detry J and Blackstone S C (1985b) Appl. Phys. Lett. 47 983

Fischer R, Chand N, Kopp W, Peng C K, Morkoç H, Gleason K R and Scheitlin D (1986a) IEEE Trans. Electron Dev. ED-33 206

Fischer R and Morkoç H. (1986b) Solid State Electronics 29 269

Fischer R, Kopp W, Gedymin J S and Morkoc H (1986c) IEEE Trans. Electron Dev. to be published.

R. Kaplan (1980) Surf. Sci. 93 145

Matthews J W and Blakeslee A E (1975) J. Crystal Growth 29 273

Metze G M, Choi H K and Tsaur B Y (1984) Appl. Phys. Lett. 45 1107

Neumann D A, Zhu X, Zabel H, Henderson T, Fischer R, Masselink W T, Klem J, Peng C K and Morkoc H (1986) J. Vacuum Sci. and Technol. B4 642

Nonaka T, Akiyama M, Kawarada Y and Kaminski K (1984) Jap. J. Appl. Phys. 23 L919

Otsuka N, Choi C, Kolodzieski L A, Gunshor R L, Fischer R, Peng C K, Morkoç H, Nakamura Y and Nagakara S (1986) Proc. of the PCSI conference Jan. 29 Pasadena CA USA J. Vac. Science and Technol in print.

Uppal P N and Kroemer H J. (1985) J. Appl. Phys. 58 2195

Inst. Phys. Conf. Ser. No. 82
Paper presented at ESSDERC 1986, Cambridge 8–11 Sept. 1986

The microstructure of the Si/SiO_2-interface: observation and related electronic properties

M. Henzler
Institut für Festkörperphysik der Universität, Appelstr. 2, 3000 Hannover, F.R.G.

Abstract. With the requirement of thinner and more perfect oxides new techniques for structural analysis have been developed. Electron microscopy (TEM), photoemission (XPS) and electron diffraction (LEED) provide information on defects, especially atomic steps at the interface. It is demonstrated by electrical measurements on different portions of the same wafer, that mobility, interface state density and fixed charge at the interface are directly related to the structural defects at the interface.

1. Introduction

The MOS-technology depends on the quality of the oxide, since the carriers in the inversion layer are very close to the Si/SiO_2 interface (some nm). For thin oxides the final interface is close to the silicon surface before oxidation. The quality of the wafer before oxidation and the process parameters during and after oxidation are therefore especially important for thin oxides. The present high standard in MOS-technology (see for example Nicollian and Brews, 1982) is largely due to trial and error, since a direct observation of the interface has been difficult. In the meantime several techniques have been developed with an atomic resolution which may be applied for interface studies. Fig. 1 shows relevant distances and some techniques for structural analysis. Whereas the optical microscope and the scanning electron microscope (SEM) have not sufficient resolution for defects important for scattering, the

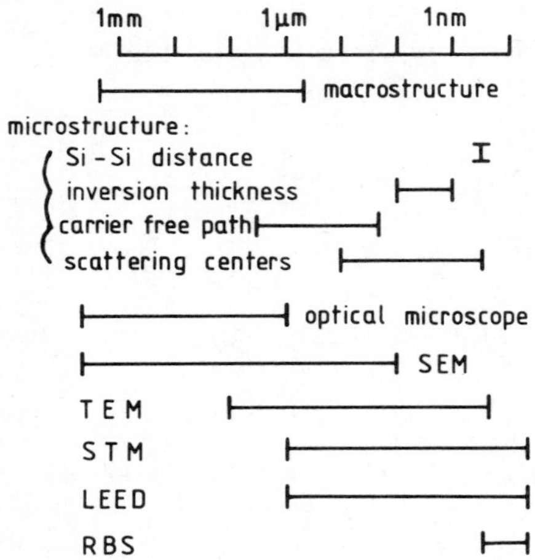

Fig. 1
The relevant distances for electron transport in devices and for some methods used for defect analysis

scanning tunnel microscope (STM) and Rutherford backscattering spectroscopy (RBS) have so far not been used for extensive Si/SiO_2-interface studies. Only three methods have recently provided systematic results with respect to process parameters and related electronic properties, and they will be described here.

2. Methods for structural analysis of the Si/SiO_2-interface

A suitable technique has to provide sufficient resolution and surface sensitivity. It should not alter or destroy the interfacial structure. Since the usual gate oxide is too thick for an investigation of the very thin interface (abrupt transition from crystalline silicon to amorphous silicon dioxide within one or at most a few monolayers), special procedures have been developed (Fig. 2). For TEM cross sections with a thickness of some 10nm have been prepared by grinding and ion milling, so that the electron beam penetrates along the interface to provide a direct image of the interface with lattice resolution (Krivanek and Mazur, 1980). For XPS the oxide is chemically thinned to less than 5nm the escape depth of photoelectrons (Grunthaner et

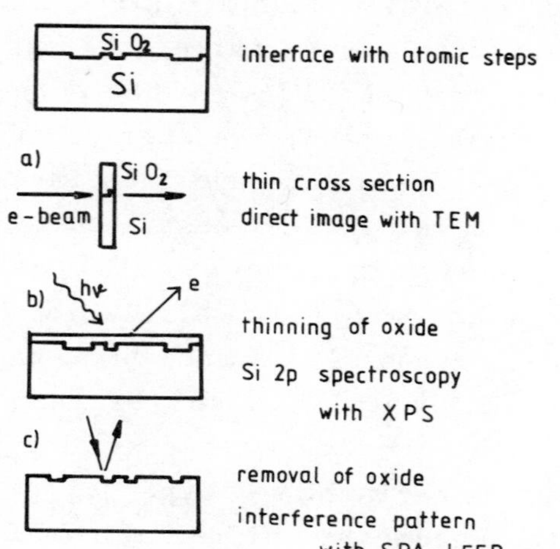

Fig. 2
Techniques for studying atomic steps and other defects at the Si/SiO_2 interface

al., 1979). For LEED the oxide is completely removed to provide an interference pattern of the bare silicon surface (Hahn and Henzler, 1981a). The kind of information from the techniques is qualitatively quite different. In TEM an image of the interface over a length of about 50nm is produced by projection along the interface over a thickness of about 15-30nm. In XPS the density of interface atoms with a special charge different from bulk (neutral Si) and SiO_2 (Si^{4+}) is derived independently from their actual arrangement. The densities of Si^{+}, Si^{2+} and Si^{3+} are used to derive indirectly the number of steps or other defects to provide sites different from the ideal interface. In LEED the structural information is within the interference pattern, which provides an average of step density, terrace width distribution, asperity height and so on.

3. Description of a real interface

An ideal interface is abrupt between oxide and crystalline silicon. The top layer of silicon is a perfect crystalline plane. For a real interface the uppermost silicon layer shows steps with perfect terraces in between. For a more accurate description any deviation from an exact lattice site should be included. Here all interface atoms of the crystalline silicon are assumed to occupy perfect sites. Therefore, a quantitative description should show the edge atom density, the terrace width distribution, the average or rms deviation D of interface atoms from an average interface and the autocorrelation of those deviations. As described by Ando et al (1982), usually a Gaussian autocorrelation $C(r)=D^2 \cdot \exp(-r^2/L^2)$ has been used for mobility calculations with D the rms deviation, r the horizontal distance between arbitrary points at the interface and L the correlation length, which is about half the average terrace width. Alternatively, an exponential autocorrelation $D^2\exp(-|r|/L)$ may be used. This corresponds to a geometric distribution of terrace widths $P(N)=p(1-p)^{N-1}$ with the probability p to find a step when going one atomic distance along the surface (Lu and Lagally, 1982). The roughness scattering of the mobility is essentially given by the product D L. (Ando et. al., 1982, Gold 1985, 1986, Lassnig and Gornik, 1986).

4. Results with TEM and XPS

Goodnick et al. (1982, 1983, 1985) have reported high resolution TEM results with a sophisticated method of evaluation. The micrographs were digitized by visual inspection to give the step edge positions individually. Due to the lattice resolution of the instrument, the minimum step height for the Si(100) is d=0.27nm. The autocorrelation of the deviation from an average inclined reference plane for a rough Si(100) interface is calculated. Its Fourier transform is given in Fig. 3 as a direct transform (FFT) and after smoothing actions (AR model). The best fit is obtained with an exponential autocorrelation. The rms deviation from a tilted plane is always close to 0.2nm, whereas the correlation length (first order background removed) varies between 2.2nm (smooth surface) and 1.0nm (rough surface). They point out that, due to limited size of photograph and projection over 10 to 30nm, the measured rms deviation may be too low by a factor of more than 2 with minor corrections to

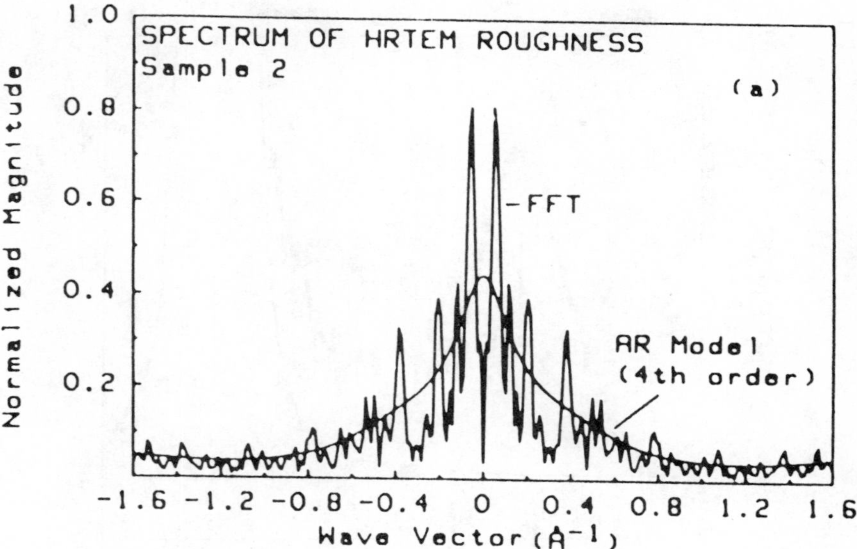

Fig. 3
Spectrum and roughness as derived from a HRTEM photograph by visual digitization and Fourier transform (FFT) and additional smoothing (AR model)
(from Goodnick et al 1985)

the correlation length. They have measured Hall mobility at low temperatures on the same wafer and could fit the measurements with a rms deviation double the value of the TEM evaluation. They also report on increased roughness due to HCl addition during oxidation.

Grunthaner et al. (1979, 1980, 1986a, 1986b) report XPS measurements. Fig. 4 shows that the energy of photo-emitted electrons out of the Si 2p level depends on the charge state of the atom. Therefore Si atoms at the interface are separated due to the charge of Si^{+}, Si^{2+}, Si^{3+} which is not found in bulk silicon or perfect SiO_2. Since the signals of Si^{+} and Si^{2+} are independant of oxide thickness from 0.7 to 2nm, those atoms are attributed to the immediate interface. As expected they find for (111) faces essentially Si^{+} and for (100) faces Si^{2+} due to the bonding configuration. If an increased amount of Si^{+} for the (100) and Si^{2+} for the (111) is found, they conclude an interface roughness due to step

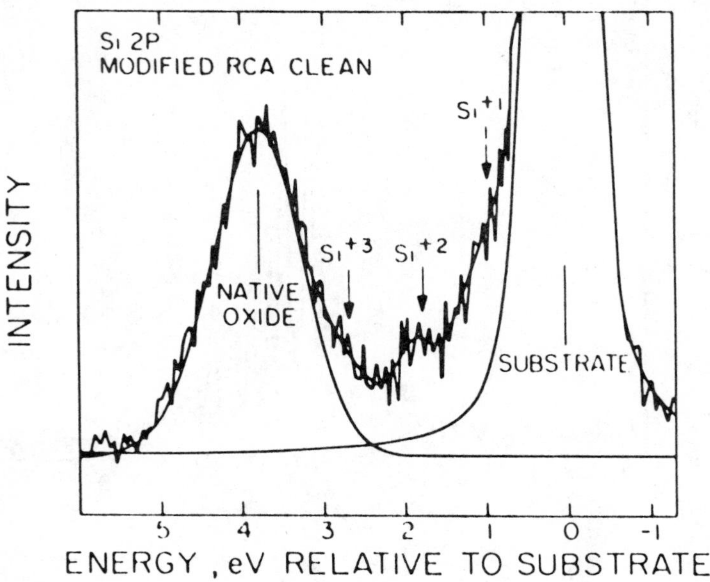

Fig. 4
Expanded Si 2p spectrum of a native oxide after chemical etching. The emission of suboxide species (Si^{+}, Si^{2+} and Si^{3+}) are indicated between the peaks from substrate and native oxide (SiO_2).
(from Grunthaner et al. 1986a)

or kink sites. They report on effects of surface preparation and of post oxidation anneal (POA). They show that the native-oxide thickness and the suboxide density and therefore defect concentration depend heavily on chemical treatment like content of H_2O_2 or H_2SO_4. They claim that the native oxide influences the following thermal oxidation. With POA at 1150°C of a (100) interface they find an increased amount of Si^{+}, which is interpreted as increased roughness. This is in contradiction to the results reported in the next section. The reason for the discrepancy ((100) or (111) face, temperature of anneal) is not known so far. They also report on radiation induced damage, which is monitored by suboxide concentrations.

5. Results with LEED

The measured intensity of a diffraction pattern contains both the atom arrangement within the unit mesh and the arrangement of the units of the surface (Henzler 1977, 1982, 1984, 1985). For the determination of atom positions extensive dynamic calculations are needed. For the arrangements of units (the only quantity of interest here) a quite simple kinematic calculation provides all needed information. The principles of evaluation are shown in Fig. 5. A surface is described by identical units consisting of one surface atom each and all underlying atoms. The integral intensity depends on the actual arrangement within the unit including multiple scattering with neighbouring units. The ratio of the actual to the integral intensity, however,

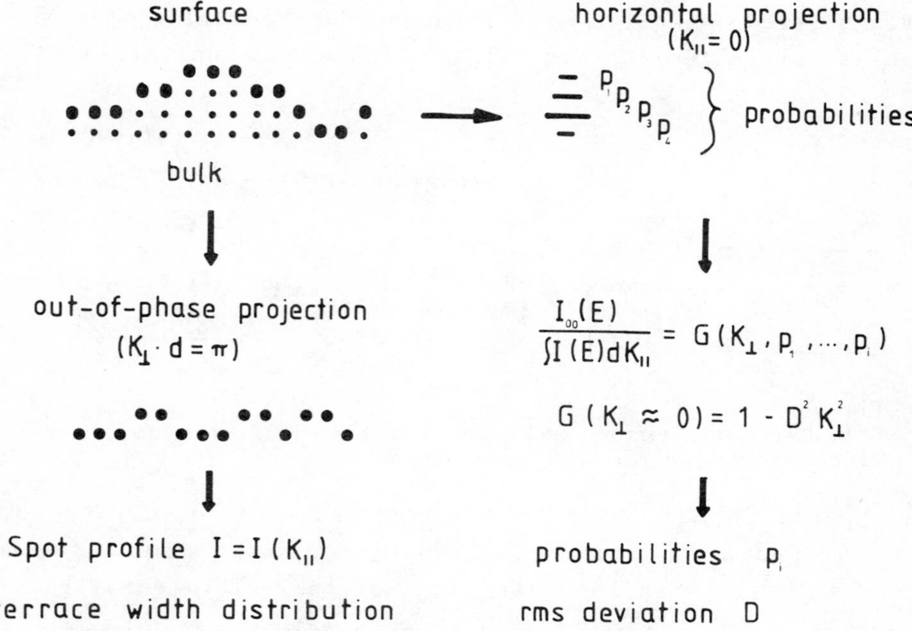

Fig. 5
Principles of spot profile analysis (SPA-LEED)
left side: out-of-phase condition ($K_{\perp} \cdot d = 2\pi(n+1/2)$) for horizontal distribution, right side: energy dependence of central spike ($K_{\parallel} \cdot a = 2\pi n$) for vertical distribution.

depends solely on the arrangements of the units, which is experimentally checked by the strict periodicity of this ratio in reciprocal space (Henzler, 1984, Horn and Henzler 1986). For anti-phase scattering of neighbouring terraces ($K_\perp \cdot d = 2\pi(n+1/2)$ the spot profile $I(K_{\parallel})$ may be directly transformed into a terrace width distribution without any fitting procedure which yields directly the step atom density (Busch and Henzler, 1986). The Fourier transform of the spot profile is the autocorrelation function. If the surface is projected horizontally, the surface is described by the probabilities of the different layers present at the surface. The lattice function $G(K_\perp) = \sum_{ij} p_i p_j \cos(K_\perp d(i-j))$ is easily calculated and easily measured. In this way the probabilities p_i and the rms deviation D from the average surface are derived. These measurements are possible only with the new high resolution instrumentation. To provide a bare oxide-free surface the oxide is etched in HF and transferred into ultra high vacuum under a droplet of methanol to avoid reoxidation (Fig. 6) (Hahn and Henzler, 1981).

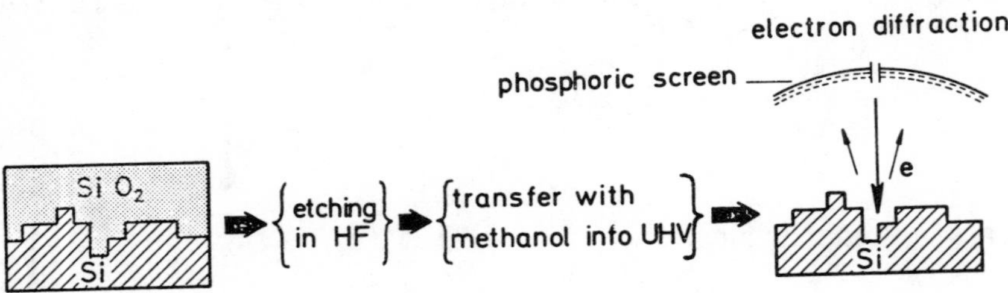

Fig. 6
Procedure for stripping and transferring a Si/SiO_2 interface into ultra high vacuum for LEED analysis.

A typical result for a spot profile of a (111) surface for out-of-phase condition is shown in Fig. 7. It consists of a central spike with the half width of the instrumental resolution and a broad shoulder, which is fitted by a Lorentzian curve corresponding to an exponential autocorrelation with coherence length 2.3nm (Busch, 1984). The relative intensity of the central spike as a function of the electron energy or scattering vector is shown in Fig. 8 for a

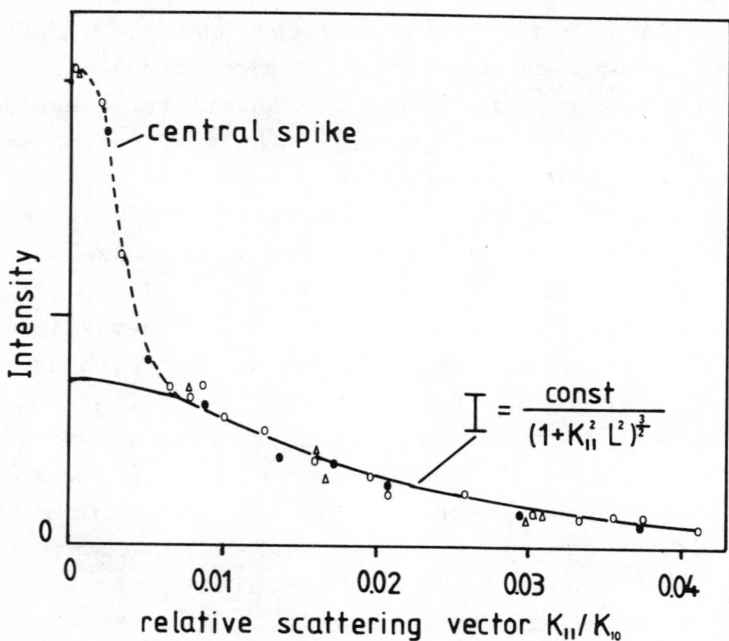

Fig. 7
Spot profile of the 00-beam of a Si(111) face at 81 and 49 eV after removal of an oxide of 30nm (1000°C, dry oxygen and annealing in N_2 at 1000°C).
(ooo 81eV, x-direction, ●●● 81eV, y-direction, ΔΔΔ 49eV, x-direction).

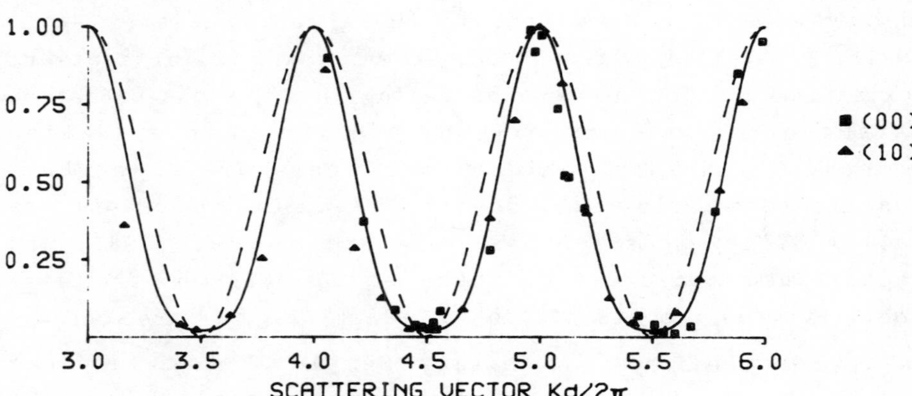

Fig. 8
Ratio of central spike intensity to total peak intensity vs. electron energy (or normal component K of scattering vector) for a Si(111) face after removal of an oxide of 5nm (900°C, dry oxygen and annealing in N_2 at 1100°C). The calculated curves are described in the text

different (111) surface (Wollschläger, 1986). It shows the typical periodic oscillations which prove the applicability of the kinematic approximation. For quantitative evaluation two calculated curves are shown. If the surface consists of two layers with equal probability, the cosine (dashed curve) is obtained. If a third layer contributes 10% and the other two 45% each, the better fit (full curve) is obtained, yielding a rms deviation of D=0.20nm. The interface consists essentially of two layers only (rms deviation close to half step height). The correlation length varies with oxidation parameters. Whereas oxidation in dry oxygen at 800°C for 1 to 35hrs always produced correlation L≈1.0-1.3nm, an annealing in nitrogen at 1000°C increases L to 2nm. For the Si(100) interface the two sublattices of the diamond lattice may appear at the surface (which are so far not distinguished with TEM). The minimum step height is 0.136nm. The number of layers contributing to the surface is much higher than for the (111) face, so that the rms deviation is larger. Very broad shoulders point to smaller correlation lengths. Precision measurements are in progress.

A large number of experiments have been performed on (111) faces with a low resolution LEED system, which provides just the average step atom density (Hahn and Henzler, 1981a, 1981b, 1983, 1984). It has been shown, that the pretreatment (oxidation and stripping or annealing in UHV), the oxidation parameters (like dry oxygen) and the post oxidation annealing (in N_2 at oxidation temperature or higher) contribute to a low step density. Of special interest are Hall mobility measurements (Hahn and Henzler, 1983) and capacitance measurements (Hahn et al., 1984). The Hall mobility measurements at liquid helium (Fig. 9) are strictly related to the step atom density. Sample R2 shows half the step density and twice the mobility of sample R1 due to annealing in N_2 with otherwise identical process parameters. Surprisingly, not only is the surface roughness scattering (high hole concentration), reduced but also the Coulomb scattering at low inversion. This suggests that there are charged centers

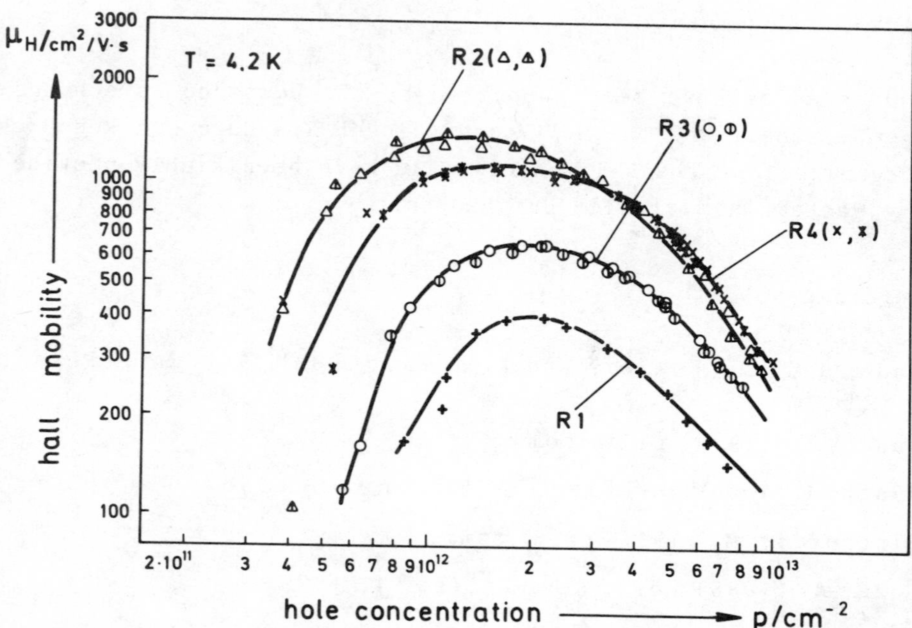

Fig. 9
Hall mobility vs. hole concentration in an inversion layer at T=4.2 K. The samples have the same gate oxide thickness but different preparation methods; dry oxidation, R2 dry oxidation and annealing in N_2, R3 wet oxidation, R4 wet oxidation and annealing in N_2 (from Hahn and Henzler, 1983).

connected with steps or kinks at step edges. The capacitance measurement showed that the fixed charge Q_f and the interface state density D_{it} is reduced due to lowering of the step density (Hahn et al., 1984). Annealing of oxides on Si(100) showed the same improvement in Q_f and D_{it} (M. Fischetti, 1984).

6. Conclusion

The different techniques provide roughly comparable results on the microstructure of the Si/SiO_2 interface, although a direct comparison of the different techniques with identical samples has not yet been tried. Electrical measurements on the same wafers demonstrate that defects like steps influence mobility and interface states. It is expected that other properties like breakdown fields and hot electron degradation depend on defects in a similar way.

Acknowledgements

Our studies have been supported by the Deutsche Forschungsgemeinschaft and the U.S. Army through its European Research Organization. The silicon crystals have been kindly provided by Wacker Chemitronic, Burghausen.

References

Ando, T., A.B. Fowler, and F. Stern (1982), Rev. Mod. Phys. 54, 437

Busch, H. (1984) priv. comm.

Busch, H. and M. Henzler (1986), Surf. Sci. 167, 534

Fischetti, M. (1984) priv. comm.

Gold, A. (1985) Phys. Rev. Letts. 54, 1079

Gold, A. and W. Götze, (1986) Phys. Rev. B33, 2495

Goodnick, S.M., R.G. Gann, D.K. Ferry, C.W. Wilmsen, and O.L. Krivanek (1982) Surf. Sci. 113, 145

Goodnick, S.M., R.G. Gann, I.R. Sites, D.K. Ferry, C.W. Wilmsen, D. Fathy, and O.L. Krivanek, (1983), J. Vac. Sci. Techn. B1, 803

Goodnick, S.M., D.K. Ferry, C.W. Wilmsen, Z. Liliental, D. Fathy, and O.L. Krivanek, (1985), Phys. Rev. B32, 817

Grunthaner, F.J., P.J. Grunthaner, R.P. Vasquez, B.F. Lewis, J. Maserjian, and A. Madhukar, (1979), J. Vac. Sci. Techn. 16, 1443

Grunthaner, P.J., R.P. Vasquez, and F.J. Grunthaner, (1980), J. Vac. Sci. Techn. 17, 1045

Grunthaner, F.J., P.J. Grunthaner, M.H. Hecht, and Di Lawson (1986a) in Proc. INFOS, eds. J.J. Simonne and J. Buxo, North Holland, Amsterdam

Grunthaner, F.J. and P.J. Grunthaner (1986b), Material Research Reports (in press)

Hahn, P.O. and M. Henzler (1981a), J. Appl. Phys. 52, 4122

Hahn, P.O. and M. Henzler (1981b), Springer Series in Electrophysics 7, INFOS, eds. M. Schulz and G. Pensl, Springer, Berlin

Hahn, P.O. and M. Henzler (1983), J. Appl. Phys. 54, 6492

Hahn, P.O. and M. Henzler (1984), J. Vac. Sci. Techn. A2, 574

Hahn, P.O., S. Yokohama, and M. Henzler (1984), Surf. Sci. 142, 545

Henzler, M. (1977) in Electron Spectroscopy for Surface Analysis ed. H. Ibach, Springer Verlag, Berlin

Henzler, M. (1982), Appl. Surf. Sci. 11/12, 450

Henzler, M. (1984), Appl. Phys. A34, 205

Henzler, M. (1985), Surf. Sci. 152/153, 963

Henzler, M. (1986), Proc. INFOS, p. 203, eds. J.J. Simonne, J. Buxo, North Holland, Amsterdam

Horn, M. and M. Henzler (1987), J. Cryst. Growth, in press

Krivanek, O.L. and J.H. Mazur (1980), Appl. Phys. Lett. 37, 392

Lassnig, R. and E. Gornik (1986) priv. comm.

Lu, T.M. and M.G. Lagally (1982), Surf. Sci. 120, 47

Nicollian, E.H. and J.R. Brews (1982), MOS Physics and Technology (Wiley, New York)

Wollschläger, J. (1986) priv. comm.

Inst. Phys. Conf. Ser. No. 82
Paper presented at ESSDERC 1986, Cambridge 8–11 Sept. 1986

X-ray lithography

A. Heuberger

Fraunhofer-Institut für Mikrostrukturtechnik, Berlin, Germany
and Technische Universität Berlin, Germany

Summary

X-ray lithography with wavelengths between 0.2 nm and 5 nm provides both high structural resolution as good as 0.1 µm and a wide scope of advantages for the application in circuit production. Examples for this better process performance compared to optical techniques are: lower particle and dust sensitivity, applicability of simple single-layer resist technique, high depth of focus without any influence of substrate material and chip topography and presumably the highest throughput of all lithography methods which are able to go into the submicron range.

However, the introduction of X-ray lithography into the semiconductor production means a revolutionary change of production technology. This begins with a completely different mask technology which makes, for example, the classical separation of mask substrate fabrication from pattern generation by different manufacturers very problematical and ends with the necessity to introduce X-ray lithography in relatively larger production capacity units consisting of a larger number of X-ray steppers. The latter is caused by the fact that a storage ring - even in the smallest version of COSY - has to supply up to 10 X-ray steppers with light in order to clearly beat the optical techniques with respect to throughput and lower cost level.

But, on the other hand, X-ray lithography provides a relatively good chance to win significant advantages in the worldwide semiconductor competition, especially for those who venture to start early with this new technology. To mention only one example, using X-ray lithography can possibly slow down the crazy spiral of future clean room demands.

To prove such statements in pilot production lines, the necessary tools and components for X-ray lithography are already or will be available for the first time on a commercial basis in the very near future. Especially steppers, sources and resists with satisfying specifications have been announced by a growing number of vendors. The most critical problem at present is the mask technology and the tools for defect elimination. However, with the existing technologies, the requirements for 0.5 µm design rules will be met very soon on a pilot scale. The first commercial suppliers of X-ray masks are now preparing for their production technologies. It seems probable that X-ray lithography can be established as a standard production tool at the beginning of the nineties.

Introduction

The future lithography requirements according to the development schedule of memory circuits and with respect to the two most critical parameters - feature size and circuit area - are demonstrated in Fig. 1 (according to [1]). This forecast is certainly speculative, at least concerning the second half of the next decade, but it demonstrates clearly the dramatic increase in the required future performance of lithography technology. It seems certain that in the first half of the nineties design-rules below 0.5 µm, chip areas of 300 mm^2, 10^9 pixels per chip and 200 mm diameter wafers must be controlled in circuit production. Furthermore, in order to minimize the pathlength in chip metallization and the problems with packaging, more chips will deviate from a quadratic shape as is already the case with memory chips (these chips are twice as long as wide). As a consequence of these

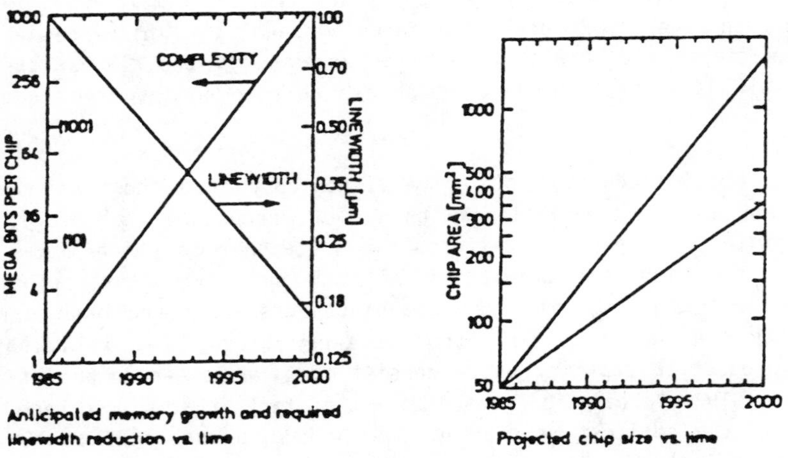

Fig. 1: Time-table in memory production (according to [1])

parameters and due to the limited image area of high-resolution optical lenses, the reduction scale of reticles for optical steppers will be further reduced. Therefore, the main advantage of optical lithography, the simpler mask fabrication with a larger scale, will vanish and the importance of techniques with 1x masks will grow in general.

Another decisive parameter is the growing number of mask levels per circuit, that is up to 15 levels and even more. From this development a very severe boundary condition arises with respect to acceptable defect density of each process or element of the complete lithography process sequence. Assuming a process consisting of nine mask levels with a tolerable lithographic defect density of $0.4\ cm^{-2}$, the defect density will be $0.009\ cm^{-2}$ per element of one level [2]. Therefore, most of the multi-layer resist processing schemes which are currently under development are very critical with respect to defects and yield. There is no doubt that a well controlled single-layer resist technique is superior to all multi-layer techniques, even though the CEL scheme, which seems to be the favourable candidate for a larger application at present, is included.

Nevertheless, very good progress in both optical and X-ray lithography in the submicron range has been made recently. The competition in future lithography techniques becomes more and more a contest between X-ray and optical lithography, both in the most advanced stage. That means, synchrotron lithography has to compete with new optical approaches such as steppers with optimized i-line lenses or deep-UV excimer lasers using new hybrid optics (glass lenses, mirrors and holographic elements) or step-and-scan mirror optics. Here, the growing number of publications concerning deep-UV lithography (e.g. very recently [3, 4, 5]) is very striking. The present situation can be characterized by the statement: the decision between optical and X-ray lithography is no longer a question of resolution capability alone but also one of technological and economical performance.

Regarding the other competing methods, comparable progress has not been made in the recent past: the situation in serial writing with electrons and ions is nearly unchanged and the former statements (e.g. [6]), that these techniques are too slow and expensive, are still valid. The deep-UV electron projection, which was very promising as a submicron full-wafer projection method according to the new Japanese approach [7] obviously cannot be developed to a step-and-repeat technique, which is necessary for larger wafer diameters. The ion projection using 1x channeling masks was studied very intensively by an American group (e.g. [8]), but recent results [9] showed very clearly that this method is limited to patterns above 0.5 μm under practical conditions, due to effects of ion penetration through the mask membrane. The very interesting demagnifying ion projection of stencil masks developed in Austria, e.g. [10], seems to be too limited in its image field for an application in standard integrated circuits of silicon. Only the 1x electron projection of stencil masks, e.g. [11], at present continues to compete with optical and X-ray lithography.

General Principles and Technical Problems of X-Ray Lithography

X-ray lithography at wavelengths between 0.2 and 5 nm is a simple one-to-one shadow-projection process with a structural resolution capability down to 0.1 μm under optimized conditions. Between mask and wafer

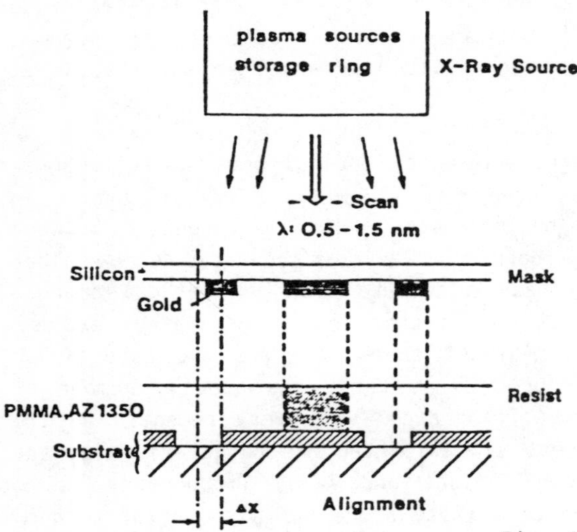

Fig. 2: Schematic of exposure arrangement for X-ray lithography using synchrotron radiation

there is a small gap, the proximity gap (typically 50 μm), which protects the mask against mechanical damage. The typical arrangement is shown in Fig. 2 which illustrates the similarity of X-ray and optical proximity projection techniques. The main factors limiting the resolution are the Fresnel diffraction, fast secondary electrons, the relatively low mask contrast attainable in the soft X-ray range, and the individual radiation characteristics of the X-ray source. However, there are two important physical facts which make X-ray lithography much more difficult than the optical process:

1. In the wavelength range of X-rays, there are no materials available which would be fully transparent in thicker dimensions (such as glass in the optical case), or which would fully absorb radiation in very thin layers (such as chrome in the optical case).

2. There are no imaging optics available to process a useful efficiency for X-rays, which means that a condensor for homogeneous illumination of the wafer is not realizable.

These two drawbacks summarize the general problems of X-ray lithography. As a result of the first, mask technology must be modified. In order to obtain a sufficiently transparent mask substrate in this wavelength region, a light element with a low atomic number and low absorption must be selected. One can obtain the required mask contrast by using a thin foil covered with a relatively thick absorber structure.

The second drawback mentioned above means that the radiation must be used in the same form (i.e., wavelength distribution and geometrical characteristics) as it is emitted from a given X-ray source. It is true that several research groups are investigating techniques for focussing soft X-rays, but the realization of a condensor with high efficiency, satisfactory stability, and application for production purposes, seems to present an insoluble problem. Therefore, the successful application of X-ray lithography depends critically on available X-ray sources.

X-Ray Masks

The key element for a successful application of X-ray lithography is a well established and controlled mask technology with respect to accuracy, stability and defect density. Considering the progress in all other elements of X-ray lithography, the successful industrial application of this new technology will depend primarily on whether the remaining problems in mask technology can be solved. Most efforts in X-ray lithography development are focussed on mask technology because this is the field where the final decision for application of X-ray lithography in device fabrication will be made.

As already mentioned, the X-ray mask consists of a thin membrane of low-Z material carrying a high-Z absorber pattern as shown in Fig. 3. The thickness of the supporting membrane depends on its mechanical stability and on both optical and X-ray transparency; it is in all important practical cases less than a few microns. On the other hand, the thickness of the absorber ranges between 1 μm and 1.5 μm in order to get a sufficiently high mask contrast. Due to the fact that the carrier membrane has a similar thickness as the absorber, the behaviour of the absorber layer severely affects the accuracy and geometrical

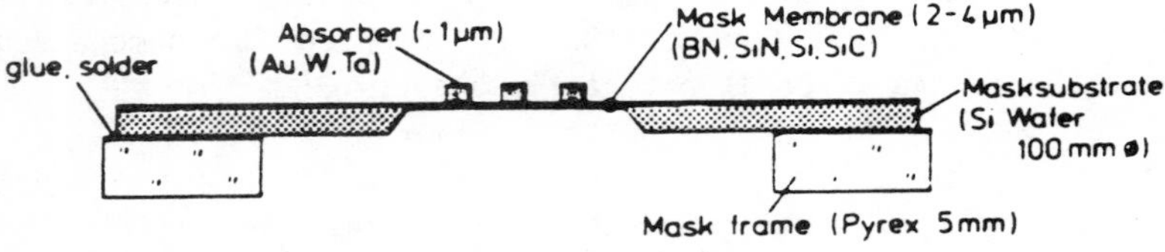

Fig. 3: Schematic cross-section through an X-ray mask

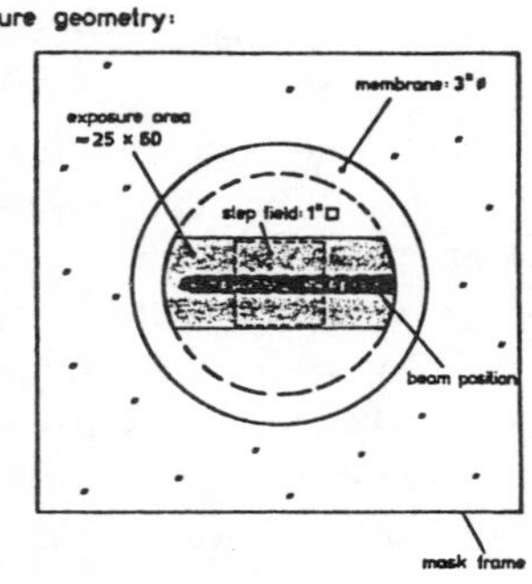

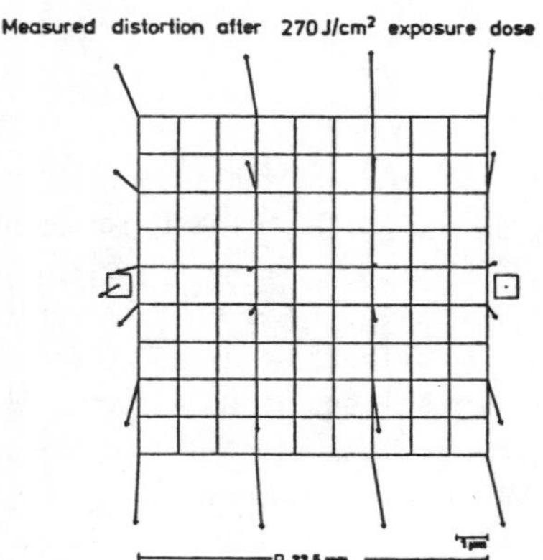

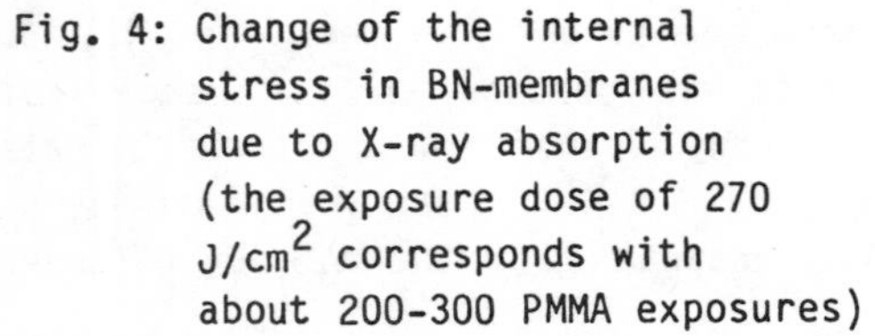

Fig. 4: Change of the internal stress in BN-membranes due to X-ray absorption (the exposure dose of 270 J/cm^2 corresponds with about 200-300 PMMA exposures)

Fig. 5: Change of optical transparency in BN-membranes applying the same exposure arrangement as in Fig. 4 (membrane thickness 4 µm, synchrotron radiation of BESSY at 754 MeV, exposure dose 500 J/cm^2) [12]

Parameter / Mask	X-ray transparency	Optical trans-parency	Young's modulus	Fabrication expenditure	Remarks
SI	good at lower energy (BESSY)	–	$2 \cdot 10^{12} \frac{dyn}{cm^2}$	+	stress engineering can be done very effectively
SIN	slightly worse than SI	++	$1.5 \cdot 10^{12} \frac{dyn}{cm^2}$	–	problems with hydrogen
SIC	comparable to SI	+	$3.5 \cdot 10^{12} \frac{dyn}{cm^2}$	?	perhaps the best distortion stability
BN	good at higher energy (COSY)	+	$2 \cdot 10^{12} \frac{dyn}{cm^2}$	–	problems with hydrogen

Table 1: Some aspects of X-ray mask materials which are considered most promising at present

stability of the entire mask. This problem can only be overcome if the stress of each layer of the mask is compensated or tuned very well. It has to be stable against all environmental influences such as exposure radiation.

Since the beginning of X-ray lithography, nearly all light elements and their compounds have been considered for masks, but only a few remain promising. Currently, the most important are silicon, silicon nitride, boron nitride, and silicon carbide. All of these materials have specific advantages and problems which are indicated in Table 1. That means that at present it cannot be decided what the final mask technology will look like. Therefore, more information is needed from a larger pilot application by fabricating test circuits.

Recent examinations of hydrogenated boron nitride membranes [12] demonstrate that radiation damage can be a fundamental wear-out for hydrogenated boron nitride films by LPCVD processes. The observed changes in both the geometrical distortion of the boron nitride films (boric oxide growths) as shown in Fig. 4 and the reduced optical transmission (Fig. 5) due to this chemical degradation make this kind of mask very problematical. The entirely inhomogeneous illumination characteristics of Fig. 4 certainly represent an extreme worst case, but these effects can severely limit the use of hydrogenated boron nitride as an X-ray mask membrane material unless they are eliminated and/or properly integrated into the exposure process. Membranes made of single-crystalline silicon seem to be extremely stable against radiation. Fig. 6 shows a measurement of a silicon membrane with a similar exposure arrangement as in Fig. 4. Up to a dose of 1350 J/cm^2 (about 1000 - 1500 PMMA exposures), no influence of the radiation on the membrane stability was observed. The distribution of the measured values represents the measurement acc-

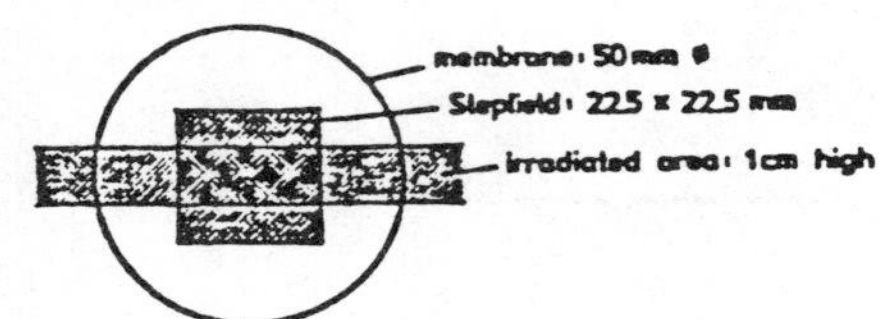

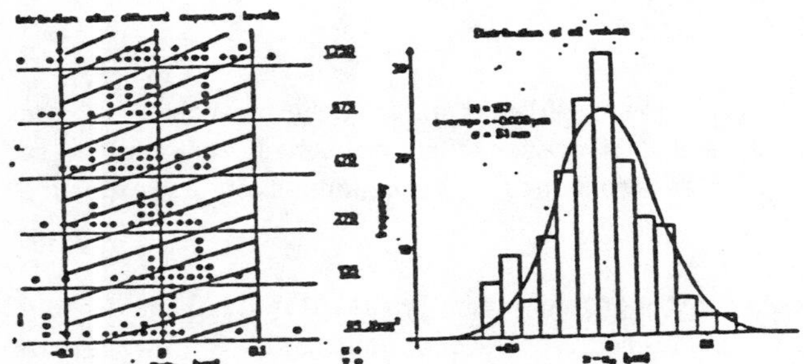

Fig. 6: Stability of monocrystalline silicon X-ray mask membranes against radiation

uracy of the used electron optical length measurement system EBMFT 5 of Cambridge [13].

Very good progress especially concerning the overlay accuracy and the distortion stability has been observed recently in the case of silicon and silicon carbide, too. SiC is the favourite material of a growing number of R.& D. groups worldwide because of its higher Young's modulus and the high optical transparency. The higher elasticity enables a reduction of the membrane thickness resulting in a higher mask contrast. In other terms, in the case of SiC a higher stress in the absorber layer can be tolerated compared to silicon. A worst case estimation of the membrane distortion due to absorber stress was given by [14]. Fig. 7 shows a summary of this consideration; as a very unfavourable situation, the chip absorber with a coverage of 50% was assumed to be located in one block on one side of the chip. Even in this case the maximum distortion for a chip area of 14 x 28 mm^2 can be held below 50 nm, if the absorber stress remains below the upper limit of $4 \cdot 10^6$ N/m^2 for silicon and $1 \cdot 10^7$ N/m^2 for SiC. For smaller absorber stresses

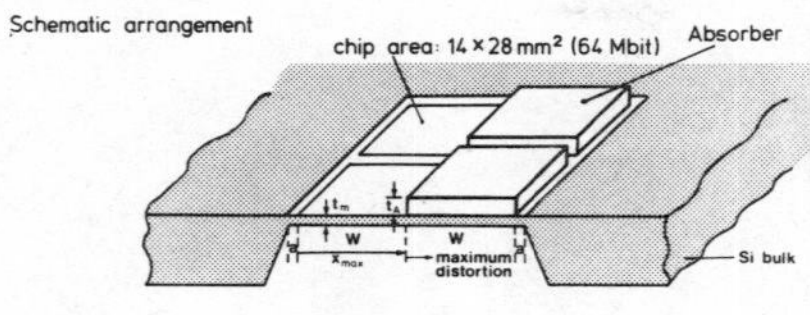

Boundary conditions

1. Absorber: 50% coverage; one block (w = 14)
2. One-dimensional treatment

$$E_M = \frac{\Delta x}{x} = \frac{\sigma_A \cdot \beta}{t_M/t_A \cdot B_M + B_A(1-\beta)}$$

σ_A: Absorber stress
$B_{A/M}$: Young's moduli
$\beta = \frac{w}{2(w+a)}$, coverage

$\Delta x_{max} = 50$ nm:

$$\Rightarrow \sigma_{A\,max} = \begin{cases} < 4 \cdot 10^6 \ \text{N/m}^2 & \text{Si} \\ < 1 \cdot 10^7 & \text{SiC} \end{cases}$$

$t_M/t_A = \frac{3}{0.8}$

Fig. 7: Mask distortion due to absorber stress (worst-case estimation [14])

the maximum distortion depends only on the chip size. In this aspect of mask technology, SiC is superior to Si.

However, due to the very widespread experience in the field of silicon technology and due to the fact that silicon is the best controlled material, silicon X-ray masks are at present in the most advanced state of development. Especially the stress engineering problem in membrane fabrication can be overcome very elegantly using epitaxial techniques by compensating the boron-induced stress by simultaneous counter-doping with germanium, as proposed by Csepregi et al. [15]. X-ray topographs taken from wafers with different levels of germanium co-doping are shown in Fig. 8. Some misfit dislocations are still observable in Fig. 8a and 8b, with a decrease in density correlated with increasing germanium concentration. The epitaxial layers in Fig. 8a and 8b have germanium concentrations of $2\text{x}10^{20}$ atom/cm^3 and $6\text{x}10^{20}$ atom/cm^3, respectively. Above a germanium co-doping level of $7\text{x}10^{20}$ atom/cm^3, the wafers are completely free of misfit dislocations. Fig. 8c shows a topograph of a wafer close to the point of complete strain compensation, with a germanium concentration of $9.8\text{x}10^{20}$ atom/cm^3. Using this technique, all stress conditions - from tensile to compressive stress - can be adjusted reproducibly in silicon membranes. A small amount of tensile stress is necessary to avoid membrane buckling. It must be tuned properly to the stress conditions of all other layers of the mask, especially to the absorber layer which is, next to the membrane, the thickest layer of an X-ray mask.

An optimized electroplating process of gold with a subsequent temperature treatment yields reproducible gold layers with a mechanical stress

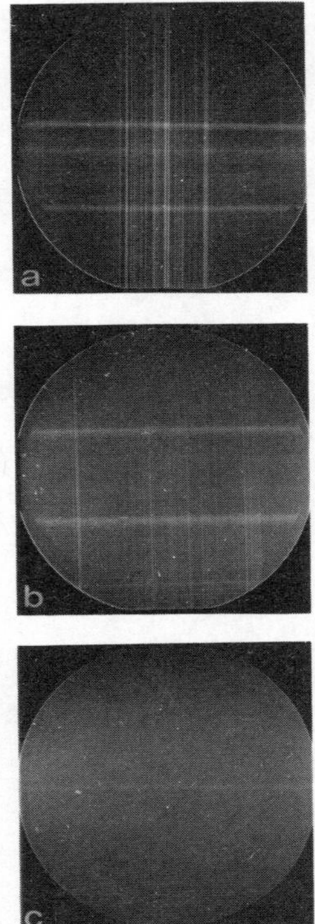

Fig. 8: X-ray topographs of epitaxial layers doped simultaneously with boron and germanium. The boron concentration is kept constant at a level of $1x10^{20}$ atom/cm^3. The germanium concentration varies from $2x10^{20}$ atom/cm^3 (a), over $6x10^{20}$ atom/cm^3 (b), to $1x10^{21}$ atom/cm^3 (c).

of less than 10^7 N/m^2, which corresponds well with the stress limits indicated in Fig. 7 [14, 17, 18]. Fig. 9 shows an example of pattern fidelity by measuring the absolute deviation of the final X-ray mask from the design data for such an optimized gold plating process [14]. The key element is the removal of the original compressive stress after electroplating, by an annealing process. The final deviation results mainly from the e-beam writing process and from the influence of the unbalanced ratio of the absorber-covered area to the free area of the silicon membrane as shown in the first drawing of Fig. 9. The measured final deviations will always remain below 0.1 μm (2 σ) within a step

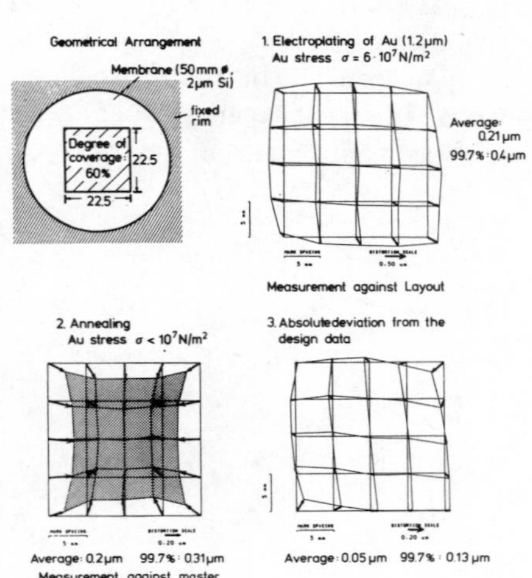

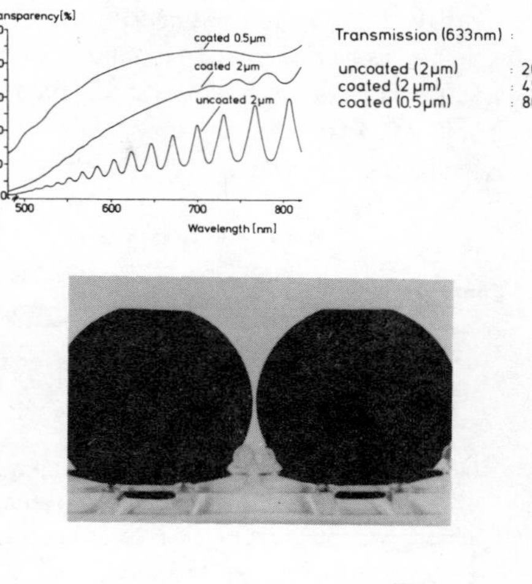

Fig. 9: Influence of the process steps on the pattern placment accuracy in X-ray mask fabrication [14, 17]

Fig. 10: Optical transparency of silicon membranes, important for alignment with optical tools

field of 25x25 mm^2 after the complete processing, if the membrane area is adjusted properly to the step field [14, 16, 17]. In respect to long-term stability of absorber layers made of electroplated gold against heavy-dose irradiation, the adjusted stress also remains unchanged within the measurement accuracy ($\pm$ $5x10^6$ N/m^2) using those doses which were mentioned earlier for the silicon membrane (10,000 exposures). Summarizing the problem of X-ray mask stability, the conclusion must be drawn that the entire mask has to be optimized in its stress behaviour. This means that all layers involved have to be tuned in respect to each other because even thinner layers can have a severe influence on the entire mask, e.g. the anti-reflection coating with SiN-layers to improve the optical transparency [16].

One of the major problems of silicon masks is the poor transparency for visible light which is required for optical alignment schemes. Membranes with a thickness of 2 µm - this is the minimum thickness for mechanical reasons - show only 20% for 633 nm. As indicated in Fig. 10 (upper part), an optical coating with a dielectric layer - e.g. both sides with SiN_xO_y, 100 nm thick - improves the transparency up to nearly 50%, which is sufficient for most practical cases. But in order to be more flexible, the combination of coatings and local thinning within the alignment mark area is certainly a good solution. The lower part of Fig. 10 demonstrates the difference in transparency of silicon membranes having a thickness of 0.5 µm and 2 µm, both coated optically.

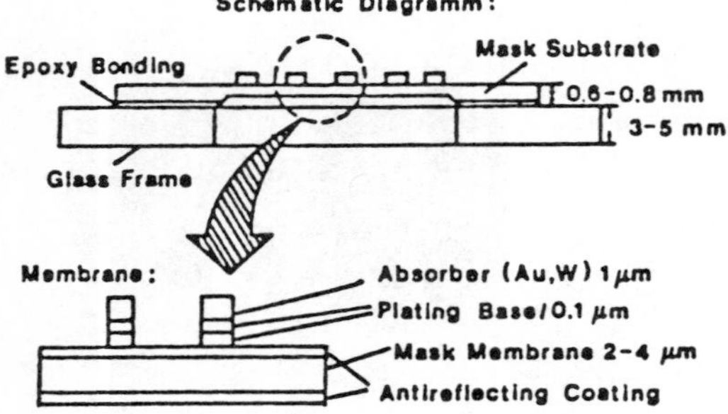

Fig. 11: Cross-section and photograph of a silicon X-ray mask with gold absorber patterns with a diameter of 100 mm and a 50 mm thin etched area [20, 17]

The cross-section and a photograph of a silicon mask including a glass-mounting frame as used presently within the cooperation in Berlin [19] is shown in Fig. 11 (preferential and self-stopping etching; layer deposition by epitaxy, CVD and plasma; batch-processing; etc. – a more detailed description of the fabrication process is given in [17]). An optimized tri-level system (PMMA/SiN/HPR 204 or other materials) structured with 50 keV electrons by using a simple proximity correction was developed for absorber formation. The present error budget in mother mask fabrication sums to a linewidth variation of ± 50 nm and is actually dominated by the e-beam writing. This kind of tri-level process reaching down to 0.2 µm with a relatively good quality is summarized in Fig. 12 and described in detail in [21]. Fig. 13 shows examples of gold patterns on a silicon X-ray mask which demonstrate that critical dimensions of 0.2 µm seem to be feasible. From this state-of-the-art in a laboratory-scale mask technology, it can certainly be extrapolated that X-ray masks can meet the requirements of a future 64 Mbit memory chip. Some target specifications of such a chip can be as follows:

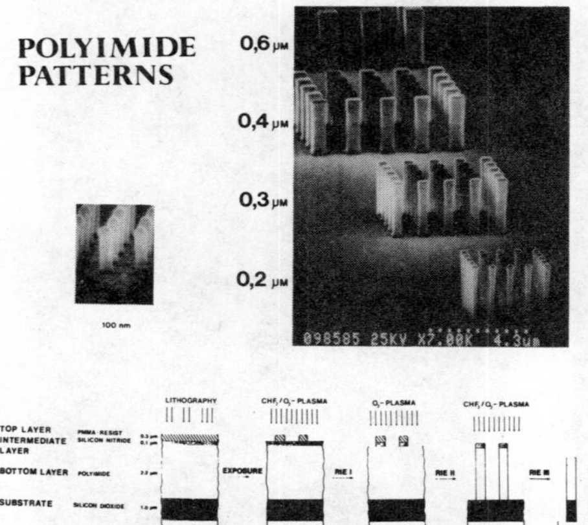

Fig. 12: Illustration of the multi-level technique used for patterning of X-ray absorbers [21]

- Pattern placement accuracy : $\leq$ 0.1 µm (3 σ)
- Step field : up to 3 x 3 cm^2
  (2 x 64 Mbit chip with area of 400 mm^2)
- Feature size : 0.3 µm / 0.6 µm pitch
- CD (critical dimension) control : ± 50 nm
- Defect density : < $10^{-2}/cm^2$
  (> 0.1 µm)
- X-ray transparency : $\geq$ 50%
- Optical transparency : $\geq$ 50%
- Absorber contrast : $\geq$ 10
- Flatness : ± 1 µm
- Radiation stability : ~ 1 ppm (10 kJ/cm^2)
- Lifetime : > 10^5 exposures

In order to reach these goals under industrial boundary conditions, further efforts are necessary, especially in the field of mask defects.

Defect Inspection and Repair

As with the problem of pattern generation, X-ray lithography has to develop its own new tools or technological methods for defect controlling, because X-ray lithography is the first method which uses 1x scale

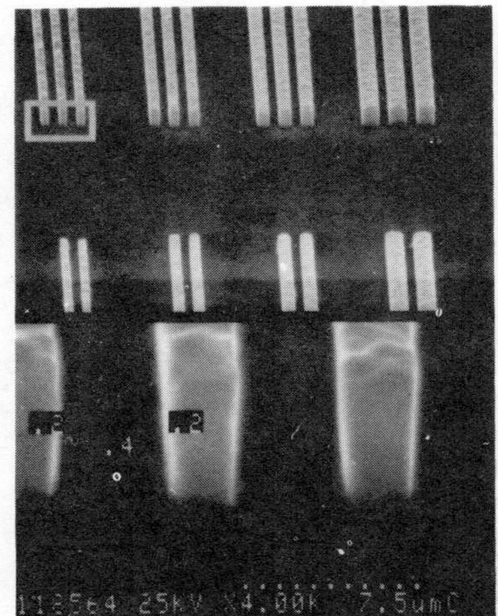

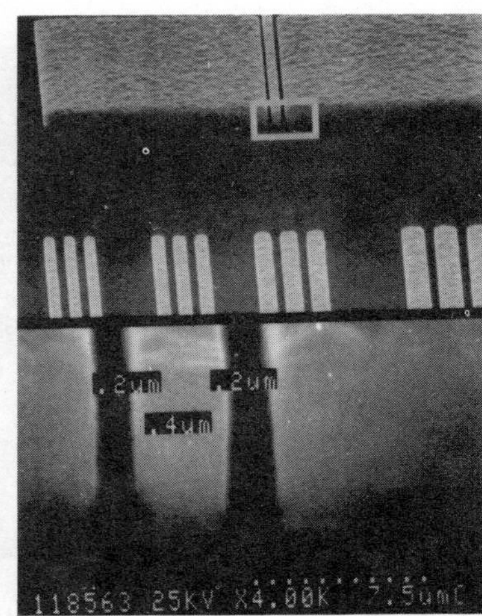

Fig. 13: Examples of critical dimensions of electroplated gold absorbers made by means of HPR-tri-level process (e-beam writing 50 keV, PMMA 0.3 µm, gold 1 µm) [14]

masks over the entire submicron range. At the present time, 0.75 µm minimum geometries can be scanned with 0.35 µm defect sensitivity with more or less satisfactory yield using existing optical equipment; improvements down to 0.5 µm geometries while retaining the same defect sensitivity seem to be possible [22]. Thus, commercially available inspection and repair techniques are not yet adequate for X-ray lithography.

Furthermore, by going down to 0.5 µm and below, simple classical rules - e.g. mask defects larger than 10% of minimum geometries are to be classified as fatal - are losing their applicability. Here the exposure behaviour of defects and particles have to be studied very carefully using process simulation and systematic experimental methods, because the influence of defects and particles cannot be scaled down linearily and there are limiting levels below which mask defects are not printed into the resist. Therefore, it is generally accepted that the defect inspection of an X-ray mask should not be carried out on the mask itself but only by studying the image transferred into a resist layer by X-ray exposure.

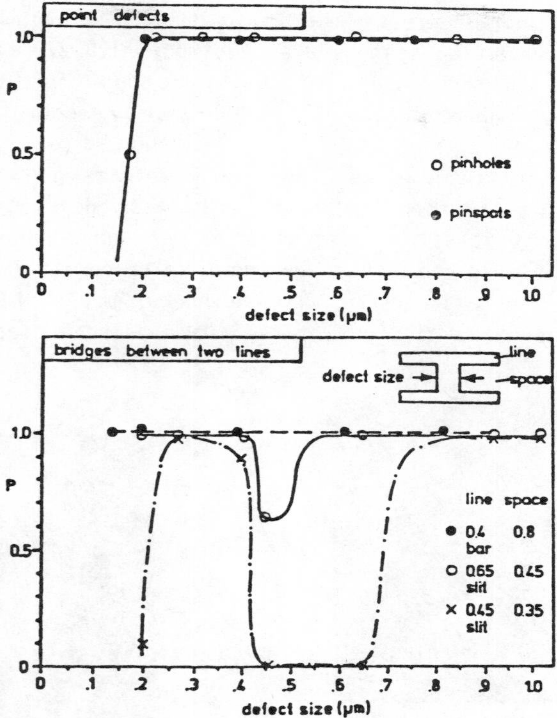

Fig. 14: Detection probability P as a function of the defect size for point defects (a) and bridges (b) obtained with an optical inspection method using the gray-level image subtraction. (Pixel size = 0.19 μm, NA = 0.95, σ = 0.39, spectrum: 3400 K, number of samples: 11) [23]

Kluwe et al. [23] carried out first studies to find the detection limits for optical techniques using the gray-level image subtraction. The detection probabilites for pinholes, pinspots and bridges achievable this way are shown in Fig. 14. Pinholes and pinspots can be detected accurately down to a size of 0.2 μm. With values below this, pixel size and optical contrast cause a steep decrease in the detection probability. It is more difficult to detect bridges between submicron lines (see Fig. 14b). They are accurately detectable between 0.5 μm bars down to a size of 0.15 μm. The same applies to slits, except for a range around 0.45 μm size.

Summarizing these results [23], a resist copy can probably be inspected optically in the case of 0.5 μm design-rules. Without optimizing the equipment, the experimental investigation of the optical gray-level subtraction method demonstrated the reliable detection of 0.2 μm point defects and bridge defects between 0.5 μm lines. Detection gaps occured at smaller linewidths. By optimizing the optical system, the detection of 0.15 µm defects in 0.5 µm structures might be possible. Below, e-beam inspection systems are required which are very problematical in respect to throughput and cost. But no information about the design of such an e-beam system is available at present.

For the repair of clear as well as opaque defects on X-ray masks, new methods including focussed ion beams and laser deposition have to be applied [24-28]. Lift-off processes for clear and laser evaporation for opaque defects fail due to the required resolution in the order of 1/10 µm and due to the required thickness around 1 µm. It has not yet been decided which method should be applied for repairing clear defects (ion

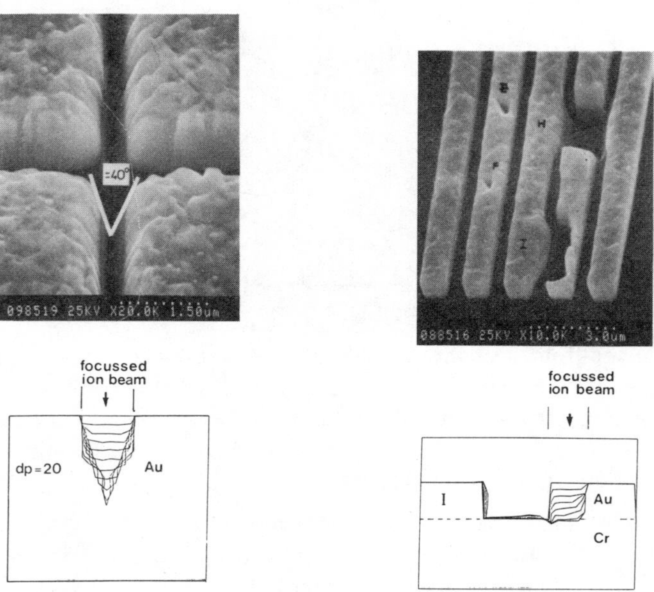

Fig. 15: Comparison between experimentally obtained V-grooves and gold redeposition after "repair" process of an X-ray mask, and a two-dimensional simulation [29, 30]

or photo-induced deposition). However, the repair mechanism for opaque defects is a straightforward application of focussed ion beams. This process step has to be applied for both kinds of defects because only the milling process with ion beams provides the required resolution. For repairing optical masks the application of focussed ion beams is not a problem as far as the opaque defects are concerned. However, in X-ray masks having absorber heights between 0.8 μm and 1 μm with aspect ratios approaching 5, the ion milling process poses some additional problems. The most important are redeposition of the sputtered material, limited maximum aspect ratios, achievable slope angle less than 90 degrees and the generation of new artificial defects. Fig. 15 shows examples of redistribution effects and the bevelling of pattern edges for the gold absorber of a silicon mask using the ion system Micrion 808 with 25 keV [29]. These results fit well to calculated resist profiles [30] (by means of an extended version of the process simulation program COMPOSITE [31]), drawn in the lower part of Fig. 15. A certain kind of "proximity effect" is created due to redeposition which significantly complicates the X-ray mask repair by focussed ions. The solution should be the application of a more complex scanning strategy combined with tilting of the mask and the use of an ion system with higher energy (e.g. 100 keV) [29]. Such a strategy is relatively time-consuming, but this is certainly not decisive because the time bottle-neck is doubtless the mask inspection.

Copying of X-Ray Masks

The fabrication of X-ray mother masks by means of e-beam writing and complicated multi-level processes is very expensive and time-consuming. This is certainly one of the most important obstacles to a practical application of X-ray lithography.

However, recent results [14, 15] show the practical feasibility of generating high quality working masks by copying mother masks using fast synchrotron exposure and simple single layer resist techniques. Applying a suitable process sequence, there is no change in the critical dimensions due to the exposure and the other technological steps like the electroplating process as demonstrated in Fig. 16 with 0.3 μm patterns. Furthermore, even mother masks, which were written without tone reversal (with the final desired coverage), can be copied in two steps by generating an intermediate negative X-ray copy with satisfactory quality. By varying the X-ray exposure parameters, it is possible to compensate for a systematic deviation of linewidth on the e-beam written mother mask from the designed values - e.g. due to an over-exposure - by the X-ray copying step. Fig. 17 shows an impressive example: two intermediate copies with different linewidths and spacings were generated from a mother mask. Second copies, which correspond very well with the original mother mask, were drawn from both.

However, such good results can only be achieved by synchrotron radia-

Resist pattern

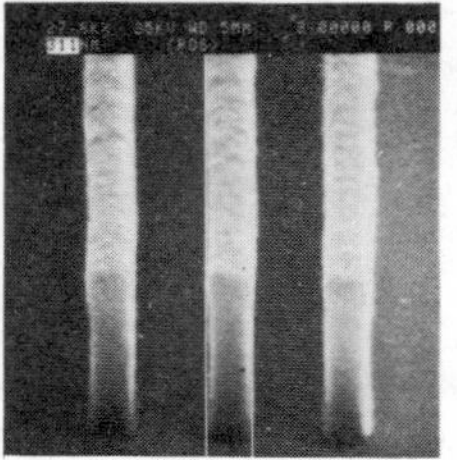

Resist : HPR 204
Exposure : Bessy 754 MeV
Resist Thickness : 1.5 µm

Absorber pattern

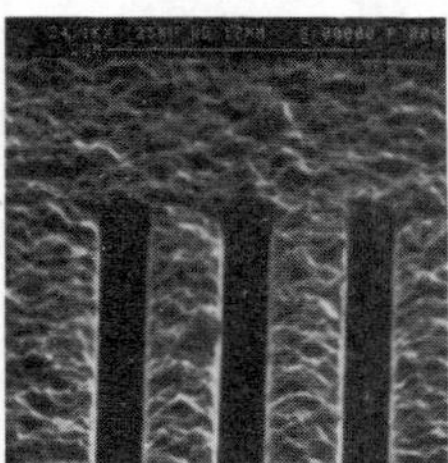

Material : Au
Thickness : 1.2 µm

Fig. 16: Example of the accurate line-width control during an optimized X-ray mask copying process using synchrotron radiation [17]

0.6 µm Steg Muttermaske 1. Kopie 0.25 µm Gräben

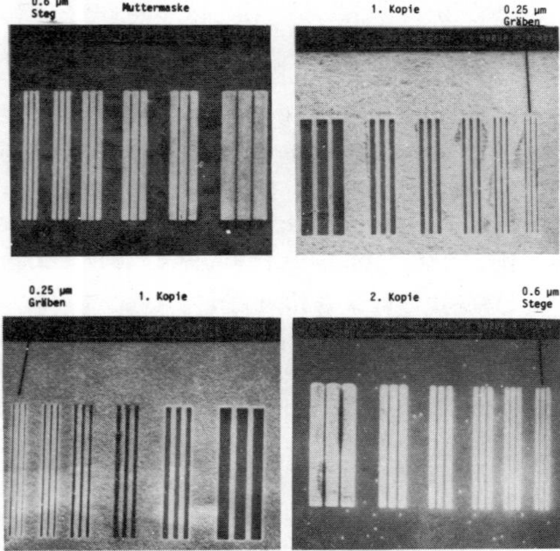

0.25 µm Gräben 1. Kopie 2. Kopie 0.6 µm Stege

Fig. 17: Demonstration of the reproducibility and adjustment of mask copying by variation of X-ray exposure parameters (from two different intermediate copies; second copies are generated which are again identical with the mother mask)

tion and an optimized copying station with sufficient heat dissipation from the exposed substrate [which is in general also a thin, etched mask membrane]. But X-ray mask-copying seems to be so successful that the philosophy in X-ray mask-making should be in future as indicated in Fig. 18.

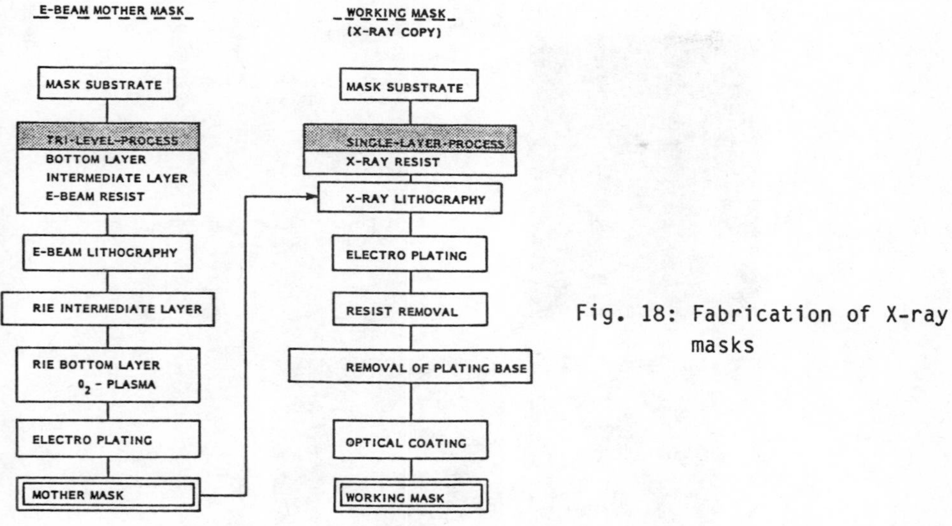

Fig. 18: Fabrication of X-ray masks

X-Ray Resists

The situation with X-ray resists is typified by the attempts to find the best compromise between X-ray sensitivity and technological stability. Nobody knows exactly what will be the final trade-off of an optimal X-ray resist in future. Therefore, depending primarily on the kind of X-ray source being used, emphasis is put either on sensitivity (X-ray tubes, plasma sources) or on stability and resolution (synchrotron radiation). Assuming the application of an advanced synchrotron source and aiming at the sub-µm range with a single resist layer technique, the requirements for X-ray resists should be as follows:

- Resolution : better than 0.1 µm
- Etching stability : at least the quality of AZ 1450
- Sensitivity : better than 100 mJ/cm^2 for high throughput (about 2 sec exposure time per step field assuming 250 mW/cm^2)
- Dissolution ratio : > 10
- Contrast : > 10

One of the best known resists for X-ray lithography is PMMA (polymer with chain scission process). Originally planned as electron-beam resist, it shows the highest resolution of all resists up to the region of 100 Å. However, the sensitivity of about 1000 mJ/cm^2 is very low and the etching stability rather poor. Therefore, many attempts were made to improve the parameters of PMMA, but they all failed. Only FBM

Exposure

Source Bessy 754 MeV
Dose 1000 mJ/cm^2

Resist

Thickness: 3 µm
Dev Time: 100 s

Substrate

1st Layer Nickel (plating base)
2nd Layer Gold 2 µm

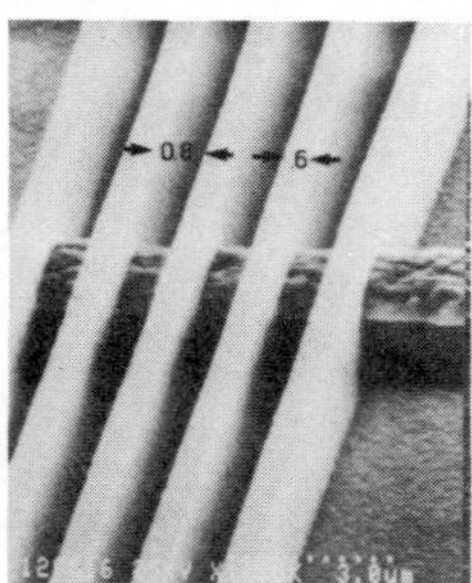

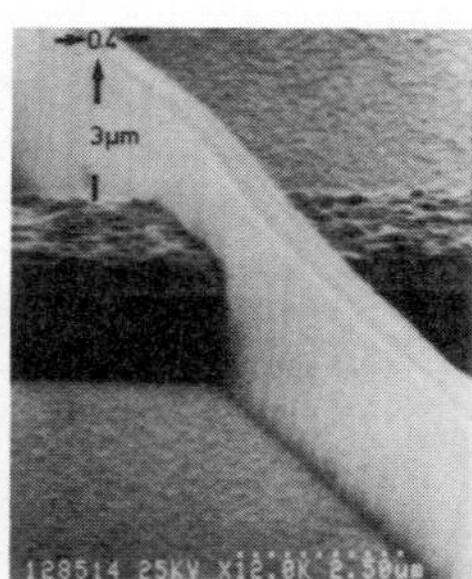

Fig. 19: Application of the diazotype resist HPR 204 with optimized resist processing for synchrotron lithography (single resist layer technique)

has some practical importance because it shows a sensitivity of about 100 mJ/cm^2 with the same stability level as PMMA.

Standard novolak-based resists (dissolution inhibitor/matrix resin) offer well proven technological stability and thus, are most desirable for a future lithography method. Unfortunately, the sensitivity towards X-rays is too low by at least a factor of 10, even for synchrotron radiation sources, if they are processed as in optical lithography. However, if the resist processing, especially the development step, is optimized for an applicaton in X-ray lithography, some optical novolak-/diazotype resists show excellent results in respect to resolution and stability even at improved sensitivity [32, 33]. The best results were obtained with HPR 204 and Hunt WX 214, which is demonstrated in Fig. 19-21. Fig. 19 shows single layer resist patterns with high resolution and aspect ratios, undisturbed by the wafer topography, made by synchrotron radiation exposure. The sensitivity was optimized and ranged around 500 mJ/cm^2 which is sufficient for medium throughput (about 10 sec exposure time per step field assuming a power density of 250 mW/cm^2). This increase in sensitivity is not combined with a loss in technological stability. Fig. 20 demonstrates the high stability against the aggressive chlorine process used for submicron aluminium patterning. Finally, the X-ray exposure is a very well controlled process using HPR 204 or WX 214, so that good pattern fidelity can be achieved; Fig. 21 shows an example of line arrays with a design measure of 0.3 µm, 0.4 µm and 0.5 µm.

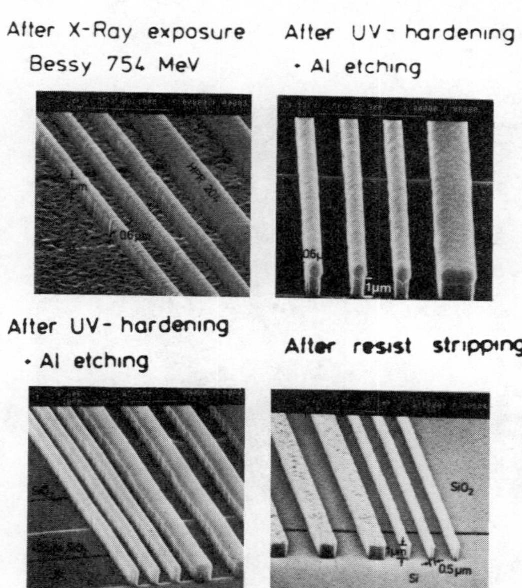

Fig. 20: Application example of HPR 204 as an X-ray resist in aluminium patterning

These resist materials are sufficient for X-ray mask fabrication and they can also serve as the first work horses for submicron device pilot processes based on X-ray lithography. But in a longer term view an improvement of the sensitivity is necessary.

To achieve higher sensitivity of novolak resists, the concept of 'chemical amplification' is introduced into an inhibitor type resist. In this newly developed resist [34], the dissolution inhibitor is destroyed via a catalytic agent generated by radiation and not directly by radiation. For the first time, a novolak-based system has been developed which meets all the requirements of a submicron-single-layer resist technology: Excellent resolution (0.25 µm lines and spaces), X-ray sensitivity well below 100 mJ/cm^2, and technological stability comparable to standard diazo-novolak resists. (Fig. 22 shows some submicron patterns made of this resist [34].)

X-Ray Sources

As initially explained, one of the most important tasks in X-ray lithography is to develop a radiation source which is capable of realizing the full potential of X-rays in lithography and which simultaneously enables economical device fabrication. There is a wide consensus that

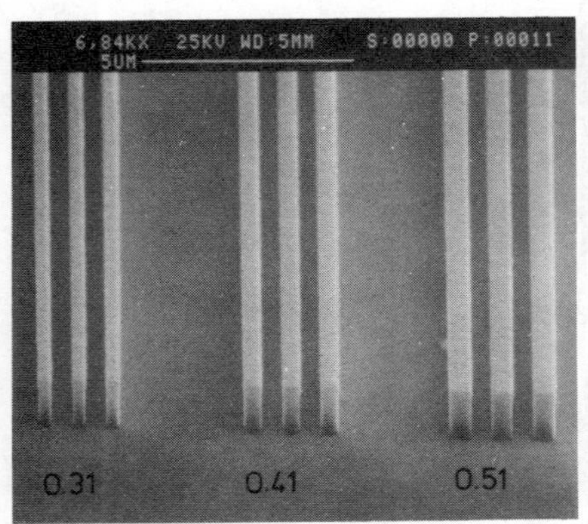

Fig. 21: Demonstration of the pattern fidelity achievable with HPR 204 and WX 214 using fine tuned resist processing and synchrotron exposure

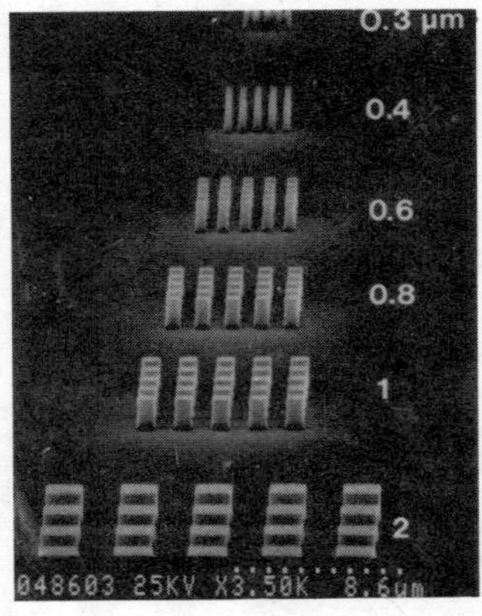

Fig. 22: New high-sensitive and process-stable novolak-based X-ray resist Resist manufacturer: Kalle/Hoechst; (BESSY 754 MeV; dose: 50 mJ/cm^2; AZ Dev.:H_2O = 1:1; Dev. time: 30 s; resist thickness: 1.5 µm)

X-ray tubes are not suitable for laboratory applications due to their poor intensity and the large focus diameter from which the radiation is emitted (blurring effect).

Plasma sources are of more interest for submicron pattern generation than X-ray tubes because of their relatively small focus in combination with higher radiation intensity, which allows increased distance between source and mask. Using plasma-source principles, laser-induced plasma and discharge plasma sources, with spot sizes down to 100um and intensity and spectral emission suitable for lithography have already been realized [35-39]. Adding an estimated lateral instability which is determined by averaging over many pulses, the effective spot size can be lower than 1 mm.

According to published results, the radiation intensity of laser-induced plasma sources are significantly lower than those of discharge plasma sources because low-cost high-power lasers are not available at the moment [35, 36]. The comparison between the different plasma sources given in [38] shows that the intensity is up to a factor of 10 higher for the best discharge plasma sources. This situation may change in future because the laser technology is progressing rapidly at present. However, at present discharge sources seem to be more appropriate for practical application.

The high X-ray brightness of discharge plasma sources is achieved by operating at much higher energy density than conventional sources. The X-rays are emitted from a high temperature plasma, which is magnetically confined and dissipates its energy isotropically. The basic principle of such sources is rather simple. An electrical current is pulsed through a cylindrical volume of gas to create a magnetic field around the gas. This magnetic field, acting as a magnetic bottle, insulates the plasma from the surrounding container walls. As indicated in Fig. 23, the magnetic shock wave moves towards the end of the inner cone where it is driven radially inward by magnetic compression to form a very high temperature plasma which emits characteristic thermal radiation in the soft X-ray range.

There is a number of alternative configurations and modifications, but two, the gas puff Z-pinch and the plasma focus, seem to be the most promising. Certainly the most advanced gas puff source for lithographic purposes [a puff of gas from a fast opening valve is utilized, which is then expanded through an annular nozzle] was recently completed at the NTT/ECL [39]. The emitted radiation power in the wavelength range of about 14 Å (Ne-filling, repetition rate 3 Hz) is sufficient to expose FBM within 20 sec using the exposure geometry of the ECL [39]. The lifetime of the electrodes is rather limited: after 10,000 shots (which corresponds with 100 PMMA exposures) the electrodes have to be exchanged.

The other promising discharge source type is a modified plasma focus (Fig. 23). The intensity of machines filled with hydrogen or deuterium, which are commonly used in fundamental research, is too weak in the soft X-ray range for application in lithography. However, an increase of the emitted X-ray power by more than two orders of magnitude has been achieved using heavy noble gases, e.g. neon or argon, for the plasma focus discharge [37]. Due to the higher mass density of the filling gas, the magnetic pressure, which acts as an accelerator of the focus plasma, has to be increased accordingly. The latter was performed by reduction of the diameter of the central electrode. The plasma is generated along the insulator at the central electrode (see Fig. 23). A sliding discharge fed by a sufficiently high electrical power flux density, produces a homogeneous and azimuthally symmetric plasma layer, which seems to be an essential precondition for an efficient X-ray

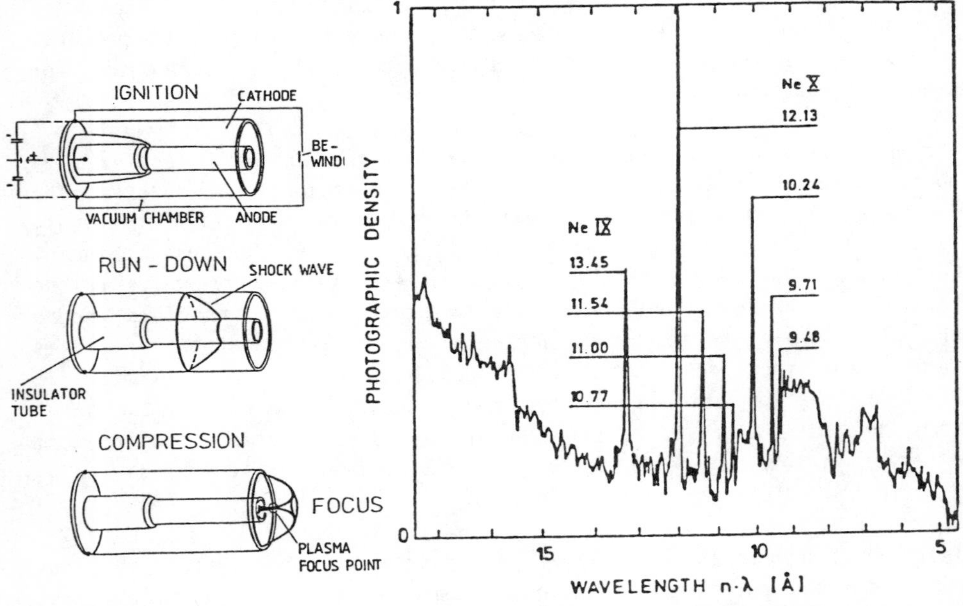

Fig. 23: Illustration of the basic principle of a plasma focus used for lithography (by courtesy of Karl Süss company)

Fig. 24: X-ray spectrum of a neon-filled plasma focus [37]

source in the final pinch phase. Fig. 24 shows a typical spectrum of a neon plasma. The spectrum is dominated by line emission of hydrogen and helium-like neon.

Sources of both types with comparable performance will presumably appear on the market by 1987. They will provide radiation power densities up to 10 mW/cm^2. The isotropic radiation, however, requires very tight control of the proximity gap in order to prevent resolution limiting run-out effects. A number of other problems are related to this kind of source, e.g.electrode erosion, debris deposition on windows, blasting shock waves limiting the window thickness, relatively soft radiation (Fig. 24), inhomogenity in time (pulse length between 20 and 200 nsec) and a high level of disturbing electro-magnetic radiation. However, these systems may close the gap between X-ray tubes and storage rings for the next few years until compact storage rings are more widely available. Furthermore, they may fill the niche in future for special application which does not require highest throughput and ultimate resolution [40].

The best radiation quality for lithography which can be achieved presently is provided by storage rings. This so-called synchrotron radiation is emitted by electrons moving with light velocity, which are deflected by a magnetic field. As indicated in Fig. 25, this radiation

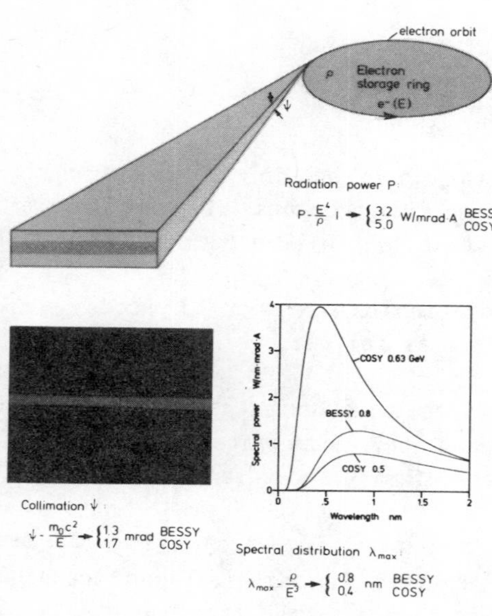

Fig. 25: Basic principle and characteristics of synchrotron radiation

is strongly collimated in a forward direction; for lithographic application it can be assumed to be parallel. The synchrotron radiation shows a wide band spectrum (Fig. 25), similar to black body radiation. The maximum output can be adjusted by changing the design parameters of the storage ring (magnetic field, electron energy, bending radius) and is not limited to some discrete characteristic lines (X-ray tubes and plasma sources). The radiation intensity is proportional to the stored current. Current up to 1000 mA can be reached at BESSY. Furthermore, the uptime of the BESSY storage ring is very high. Fig. 26 shows the beam-time of BESSY in hours per year since starting the machine.

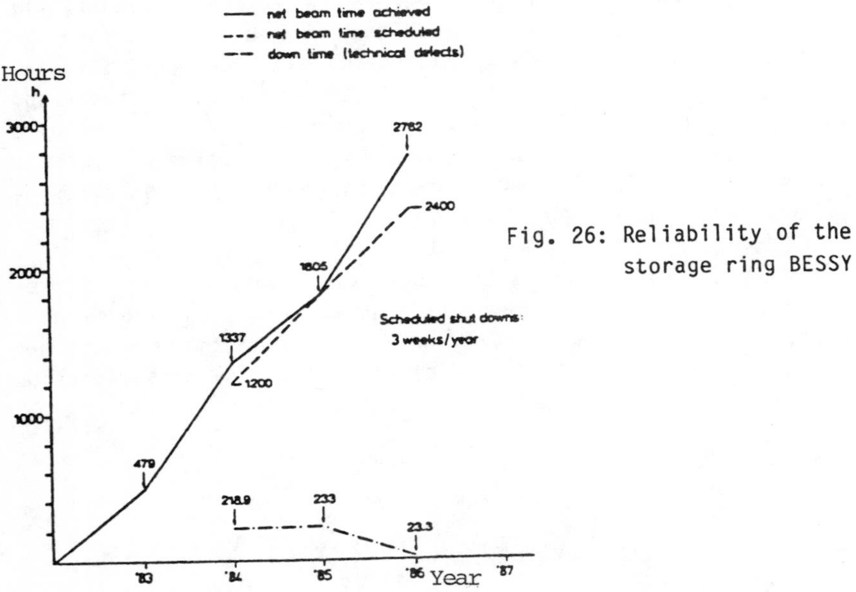

Fig. 26: Reliability of the storage ring BESSY

The continuous growth is due to an increase in the number of shifts. As a consequence of some improvements of the storage ring during the initial phase, the uptime in 1986 will presumably exceed 95%.

The advantages of synchrotron radiation with respect to process performance for semiconductor fabrication are as follows:

- Parallel radiation which provides X-ray exposure with large depth of focus (several 10s μm) and without any geometrical distortion (blurring, run out) even in the case of mask and wafer unevenness

- High intensity up to several 100 mW/cm^2 which allows the use of medium-sensitive but stable resists for single-layer technique in combination with short exposure time and high throughput

- Broadband emission which lowers the Fresnel diffraction, the resolution limiting effect in X-ray lithography

- Quasi-uniformity in time which is a precondition for optimal heat dissipation from the mask during exposure.

Based on these facts, by now there is a consensus worldwide that only synchrotron radiation allows a full realization of the advantages of X-ray lithography in semiconductor production.

Naturally, there are also some problems related to an application of storage rings in lithography. The main problems result from the huge size of such machines and from the inhomogeneous radiation characteristics in the direction normal to the orbit plane as shown in Fig. 25. The latter can be assumed to be solved because three experimentally proved solutions are available. Two of them are illustrated in Fig. 27. First, the application of scanning grazing-incidence mirrors was successfully demonstrated [42]. In this case, a plane mirror made of Zerodure size of 15x100x3 cm^3 and a surface roughness of <5 nm RMS was used. During the exposure, the mirror is scanned with a frequency between 0.1 and 1 Hz: The scanning range amounts to 0.8° ± 0.3° with a dose uniformity in the resist better than 10%. The mirror is kept under a vacuum

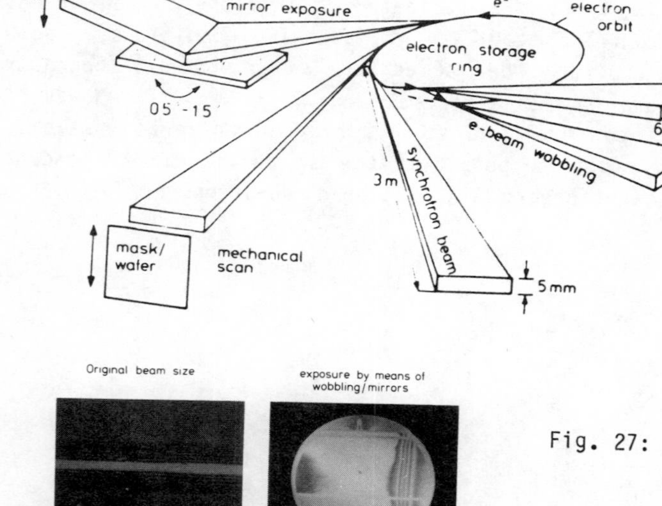

Fig. 27: Large-area exposure using synchrotron radiation

better than 10^{-9} mbar and thus the lifetime exceeds 10,000 PMMA exposures. With this method, the exposure of 6x6 cm^2 stepfields is possible but with an intensity loss of nearly 50%. Furthermore, a mirror chamber would be the most costly part of a beam line.

An alternative method is the stimulation of electron oscillations in the storage ring as also shown in Fig. 27. The feasibilty of this technique up to 6 cm high step-fields was initially demonstrated at BESSY [41]. The oscillation frequency could be varied between 0.1 and 1 Hz. The developed resist profile in the lower part of Fig. 27 reveals an homogeneous exposure over the full angle range. Comparing both methods, it turns out that electron beam wobbling is preferable because there is no intensity loss and there are no moving parts. The main disadvantage, however, is the required larger vacuum aperture which can increase the cost of the storage ring appreciably. Therefore, the mechanical scanning of the aligner mask/wafer set-up, whose feasibility as already been demonstrated, might be the most economical solution.

Compact Storage Ring

Synchrotron sources operated at present are unacceptable for utilization in semiconductor fabrication due to their large physical dimensions and high cost. Such machines have numerous options for fundamental research problems. These features are unnecessary in a dedicated lithography storage ring; e.g. high brilliance, extreme stability of focussing conditions, precision current control, etc. In order to achieve an optimized lithography storage ring, there are several alternatives. However, to develop a storage ring as cheaply as possible and simultaneously as small as possible, the so called "COSY"-concept appears to be a straightforward and suitable solution. In Fig. 28 a

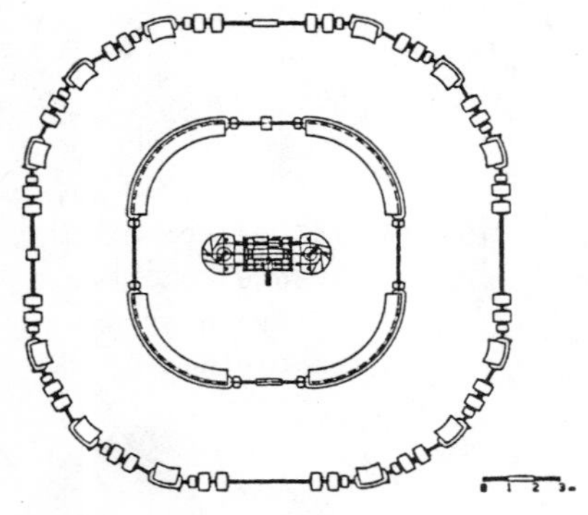

Fig. 28: Comparison between the BESSY storage ring, a low-cost conventional ring with 1 T magnets and COSY

Fig. 29: Assembly model of a future COSY lithography station (by courtesy of COSY MicroTec)

comparison of the size of BESSY, a cost-optimized normal conducting conventional storage ring with 1 T magnetic field, and the superconducting COSY can be seen. The final size of COSY is no larger than a high current ion-implanter because power supplies and injector can be installed remote from the clean room area. A model showing how the COSY-assembly might be arranged with beamlines and steppers is shown in Fig. 29.

The design of the first prototype of COSY, which is presently being developed in Berlin [43], uses a pair of superconducting magnets with a nominal induction of 4.47 T on the particle orbit. This choice reduced the particle energy to 592 MeV and the bending radius to 0.44 m for a critical wavelength of 12 Å. The orbit has a race-track shape with two straight sections of 3.4 m length as shown in Fig. 29. The circumference is 9.6 m and the area needed by the storage ring itself is only 2x5 m^2. The overall area including the injector, rf-transmitters, power supplies and He-liquefier is about 115 m^2. Each magnet consists of 3 pairs of banana-shaped coils, with a bending angle of 180°. The design orbit lies in the magnetic midplane which is defined by a symmetrical arrangement of each pair of coils above and below this plane. Some important data of COSY are:

- Energy	:	600 MeV
- Bending radius	:	0.44 m
- Wavelength of maximum intensity	:	0.5 nm
- Power on wafer (3 m, 10 µm Be, 2 µm Si)	:	~ 100-200 mW/cm^2
- Lifetime	:	~ 6 hours
- Weight	:	8000 kg
- Power consumption	:	300 kW
- Injection energy	:	50 MeV (Race-track microtron)
- Ramping time	:	3 min (50-600 MeV)

More details about COSY are given in [43].

To avoid an expensive and large "full energy injector", it is advantageous to use commercially available accelerators in the 10 MeV - 50 MeV range like linacs or microtrons. The electrons then have to be accelerated in the storage ring to their final energy by fast magnetic ramping. Related to this "low energy injection" are a number of severe technical problems of to beam instabilities, which represent the main development risk of COSY. Some publications, e.g. [44], predicted that the COSY-concept would not overcome these problems. Therefore, a test COSY with normal conducting coils but identical to the final COSY prototype in other respects was built to study the low energy injection in detail . Fig. 30 shows a photograph of this test version of COSY. A current of 25 mA with a lifetime of several minutes, which was higher than expected, could be accumulated very easily. Extrapolating these first experimental results, it appears that the specification parameters of the first COSY prototype, which will be in operation within the first 6 months of 1987, will be reached.

Similar progress could be achieved in developing a simple low cost beam-line suitable for COSY. The role of a beam-line is complex, including:

- transfer of the radiation from the storage ring at 10^{-11} mbar to the X-ray stepper in normal atmosphere

- high transparency in the usable wavelength range and absorption of the radiation which causes mask heating

- prevention of filling the storage ring with air in case of a window breaking.

Fig. 30: Normal conducting test version of COSY (by courtesy of BESSY and MicroTec)

Former beam-line concepts were very expensive and complicated, especially due to many UHV components. The new COSY beam-line concept (Fig. 31) using a silicon membrane to separate the UHV part from the coarse

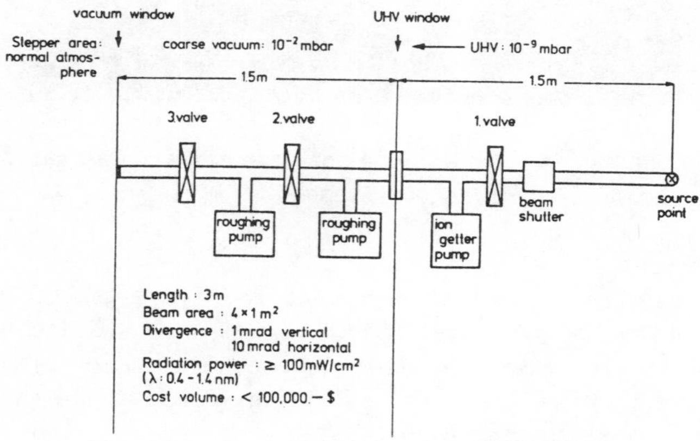

Fig. 31: Beam-line concept for the compact storage ring COSY

vacuum part, is a simple set-up with only 3 m in length as a minimum. That means, the length of beam-lines will only be determined by the geometrical size of the steppers which have to be arranged around COSY.

X-Ray Stepper

The first X-ray mask aligners for laboratory applications have become available commercially, but they are only suitable for X-ray tubes or plasma sources. The first known prototypes of X-ray steppers with vertically positioned mask and wafer, which are necessary for operating with synchrotron radiation, were built by IBM (Brookhaven) [45] and the cooperating group in Berlin (Süss, Siemens, IMT) [46]. The steppers which have been operating for two years in Berlin, (Fig. 32 shows a

Fig. 32: Photograph of an X-ray stepper of the first generation in Berlin (by courtesy of Süss and Siemens)

photograph of the stepper mechanics) represent the first generation of a commercially oriented stepper development by Süss. The primary task of the first generation was to study the achievable alignment accuracy under realistic conditions and to learn more about handling X-ray masks in a stepping process.

Alignment is performed by a pattern-recognition process, which was developed by Siemens [47], using an optical method which is fairly insensitive to process-induced changes of the wafer alignment markers. Repeatability measurements of detection and positioning accuracies show very attractive values with a 1 σ at about 5 nm, rather independent of resist thickness, defocus, and edge angle of the markers.

To study the alignment accuracy systematically, the method of double exposure of a wafer with the same mask - illustrated in Fig. 33 upper part - was used. After the first exposure, the mask was shifted a short

Principle of double exposure:

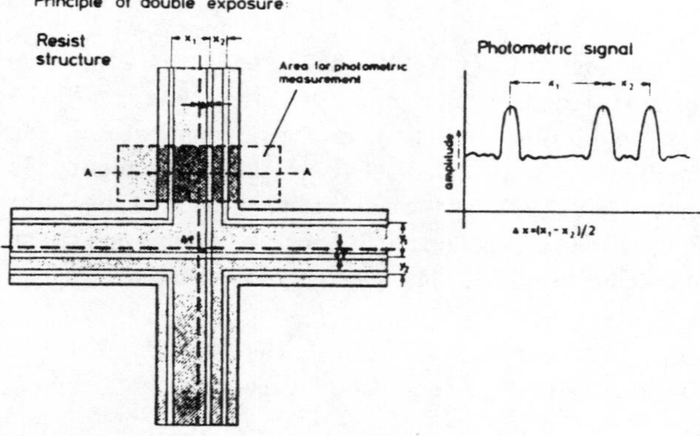

Experimental results:
Exposure system: MAX-1 K. Suss
Mask: Silicon (2 μm) , Marker material: SiO_2 , Resist: FBM 120

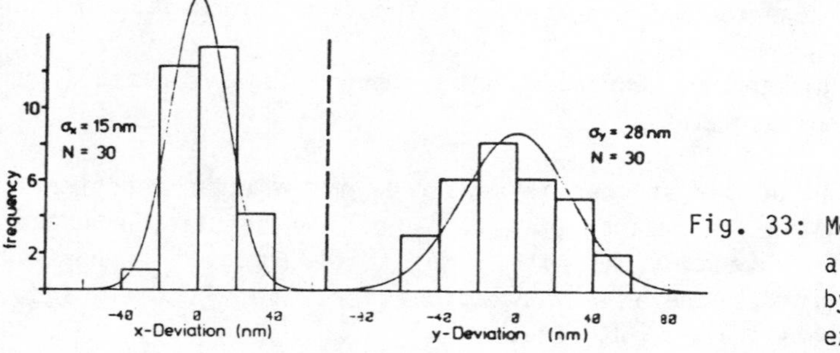

Fig. 33: Measurement of the alignment accuracy by means of double exposure

distance in order to place the single cross in the center of the double cross which was generated by the first exposure. By this means, the alignment accuracy without mask and wafer distortions can be measured. The results of the double exposure experiments indicate that an alignment accuracy better than 0.1 μm can be achieved. That means that

- the alignment method is able to determine the relation of mask to wafer at a proximity gap up to 50 μm in the range of 30 nm (Fig. 33 lower part)

- the feedback loop of the piezo stage is able to maintain the mask and wafer position in the range of 30 nm during exposure.

The actual overlay accuracy of the different mask levels after exposure have been determined by using special overlay test patterns. The results reveal a total overlay accuracy including misalignment, exposure induced distortions etc. of 0.15 µm (2 σ) for a stepfield of 25x25 mm^2. Due to the exposure arrangement in normal atmosphere, no distortions caused by heat dissipation problems from the mask could be observed. This agrees well with previous simulations and measurements [48].

The above described first prototypes (MAX I) have only a very low working speed. Now a new generation is under construction (MAX 3 or XRS 200) which is based on the experience with the first generation and optimized for high throughput. It is planned that the first prototypes will be ready at the beginning of 1987 and then commercialized by Karl Süss. A sketch of the most important parts - wafer and mask handling, stepping unit - is shown in Fig. 34; the main features are:

- MAX 3 is designed as a scanner, i.e. the X-ray beam is stationary and the aligned mask and wafer "package" is scanned through the X-ray beam.

- the mask is carried by a six-axis piezo stage. It allows one to adjust the mask plane perpendicular to the X-ray beam (3-z-piezos) and to move the mask in x, y_1, y_2 for fine alignment.

- during fine alignment and exposure, the stepper stage is clamped by vacuum to a granite plate.

- the x,y-position of the stepper is measured with a laser interferometer. This allows exact alignment of the stepfield on the wafer so that mix-and-match with a full field aligner is possible. It also reduces the x,-y corrections for fine alignment and, therefore, decreases alignment time.

- to speed up the wafer change and make the wafer transport more reliable, the complete chuck is taken out of the stepper stage and replaced by a second one carrying the next prealigned wafer. The wafer handler uses a pick-and-place system for wafer and mask transport.

The throughput can be calculated by the following parameters: Overhead

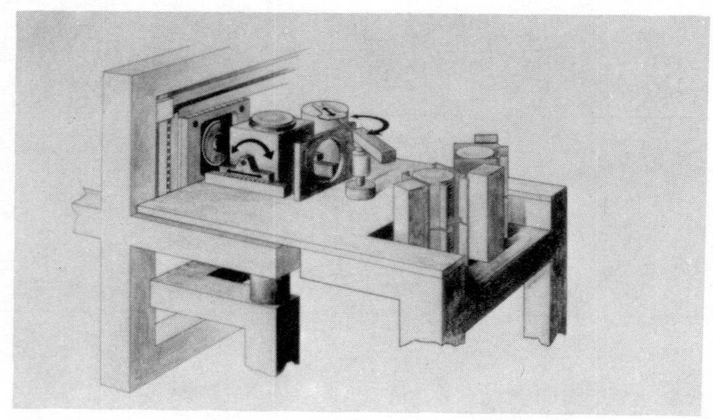

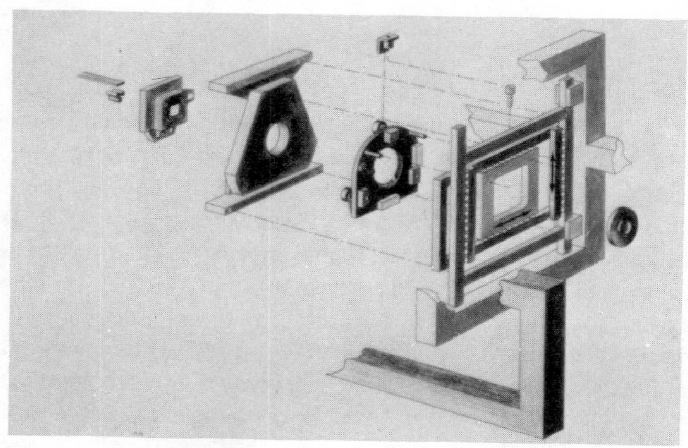

Fig. 34: Advanced X-ray stepper for synchrotron radiation with high throughput
upper part: wafer and mask handler; lower part: stepping and scanning unit (by courtesy of Karl Süss)

time for wafer change 14 sec (load/unload 12 sec, course alignment for the first field 2 sec); and machine time for one step field 2.5 sec (stepping 1 sec, proximity 0.5 sec, alignment 1 sec). These parameters yield the wafer throughput given in Table 2.

Wafer diameter (mm)	Stepfield size (mm^2)	No. of Stepfields	Throughput (Wafer/h) Exposure time (s) 0	1	4
200	25 x 25	44	29	21	12
200	36 x 36	21	54	41	24
150	25 x 25	32	38	29	16
150	36 x 36	16	67	51	31

Table 2: Wafer throughput of the X-ray stepper MAX 3

Assuming COSY and the new high sensitive novolak resist described before, an exposure time of 1 sec can be achieved.

Conclusion

Systematic investigations of the X-ray lithography application limits show that this technique can be utilized under reasonable preconditions down to pattern dimensions of 0.2 µm. The experimental results correspond well with calculated resolution limits by three-dimensional process simulation programs [49]. (An illustrative example is shown in Fig. 35. The upper part is the SEM photograph of a resist test pattern which was designed to study the resolution limits in the cases of lines, bridges, posts, gaps and so on; the lower part shows the corresponding simulation results. From this study the conclusion can be drawn that lines and bridges can be generated by synchrotron lithography down to 0.1 µm, posts down to 0.2 µm and gaps down to 0.3 µm. By reducing the proximity gap of 50 µm to e.g. 30 µm, the limits can even be improved.)

The low particle sensitivity of X-ray lithography is of the same importance as resolution capability. Organic and silicon particles disturb the resist image at thicknesses larger than 3.4 - 4 µm. From a size of 0.3 µm onwards, gold and iron particles already disturb. Slightly more severe conditions must be applied for particles covering structures because of linewidth variations, but all in all this is an exorbitant insensitivity against dust particles. Together with the other advantages of synchrotron lithography - like single layer resist technology with high throughput, extreme high depth of focus etc. - this technique should doubtless be the favourable technology for the submicron range. Furthermore, the present state-of-the-art seems to be very promising, so that X-ray lithography will be ready for circuit production at the end of this decade. The equipment will be available because many companies and new enterprises have announced their first serial products.

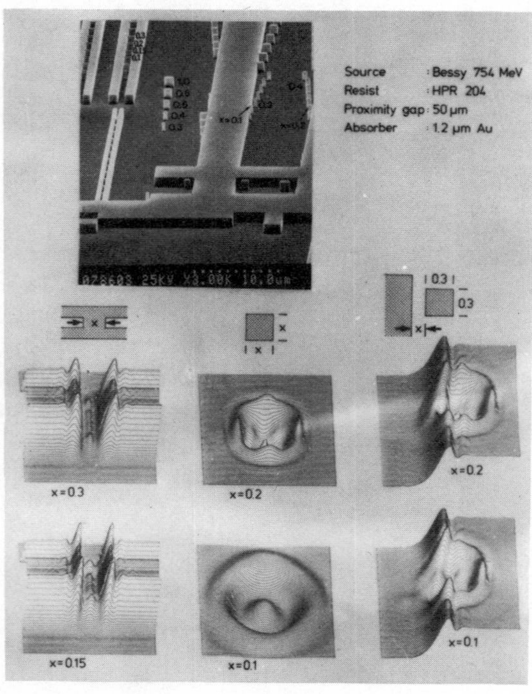

Fig. 35: Resolution limits in synchrotron lithography (comparison between exposed patterns and calculated profiles by means of XMAS)

In spite of these favourable circumstances, the present situation is characterized by a general uncertainty in the evaluation of X-ray lithography. The reason for this has nothing to do with storage rings, the key problem is the fact that many specialists worldwide do not believe that a satisfactory 1x scale mask technology can be established in the submicron range. Therefore, there will be no commonly accepted strategy to use X-ray lithography: each company has its own often contradictory philosophy. The most extreme approach under discussion is to use the benefits of X-ray lithography for the uncritical levels of future circuits and advanced optical lithography for the critical high resolution levels by accepting chip composing.

The only consequence of that discussion can be that the efforts in X-ray lithography should be focussed on the goal to demonstrate the feasibility of an industrial X-ray mask technology for highest resolution.

The main application of X-ray lithography is certainly the field of microelectronics. But there is a new fast-growing discipline called "micromechanics" which means the technology for the integration of entire systems consisting of sensors, logic and actuators [50]. These integrated systems generally need three dimensional patterns with a significant extension in Z-direction which means high aspect ratios. X-ray lithography using synchrotron radiation is the only three dimensional lithography technique; patterns as shown in Fig. 36 can only be generated with this method. This gives scope for further applications for X-ray lithography using synchrotron radiation.

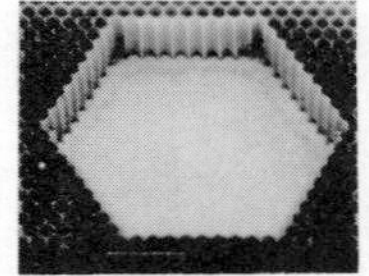

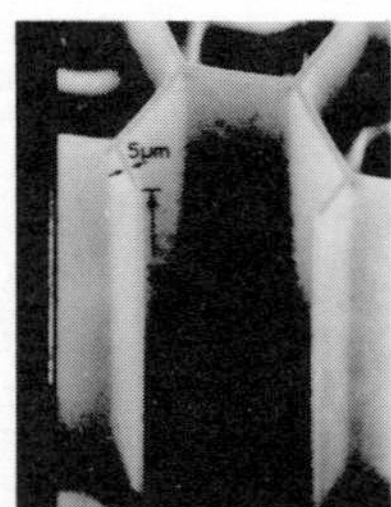

Fig. 36: Exposure of very thick resist layers using synchrotron radiation for micromechanical application (by courtesy of KFK Karlsruhe)

Acknowledgement

I wish to thank H. Betz for helpful discussions and suggestions. I also wish to thank Mrs Heid, Mrs Richters, Mrs Sibaei, Mrs Wilhelm, Mrs Grandorff and Mrs Sponholz for their support in preparing the manuscript and the figures.

References

[1] A.D. Wilson, SPIE, 1985, St. Clara.
[2] J. Lee, Solid State Technology, p.143, (1986).
[3] R.T. Kerth, K. Jain and M.R. Latta, IEEE Eletron Dev. Lett. EDL-7, 299. (1986).
[4] D.J. Ehrlich and M. Rothschild, 30th Intern. Symp. on Electron, Ion and Photon Beams, Boston (1986).
[5] F.N. Goodall, R.A. Moody and W.T. Welford, Optics Communications 57, 277, (1986).
[6] A. Broers, Microcircuit Engineering '84, ed. A. Heuberger and H. Beneking, Academic Press, p.3, (1985).
[7] I. Mori et al., Microelectronic Engineering 3, 69 (1985).
[8] J. Bartelt et al., Electrochem. Soc. Ext. Abstracts, 82-1, 454 (1982).
[9] L. Csepregi and A. Heuberger, to be published.
[10] W. Maurer, G. Stengl, F. Rüdenauer, H. Löschner and W. Fallmann, Microelectronic Engineering 3, 167 (1985).
[11] H. Bohlen and W. Kulcke et al., IBM J. Res. Dev. 26, 568 (1982).
[12] W.A. Johnson, R.A. Levy, D.J. Resnick, T.E. Saunders, A.W. Yanof, H. Betz, H.L. Huber and H. Oertel, Int. Conf. on Electron, Ion and Photon, Beams, Boston (1986).
[13] W.H. Brünger, H. Betz, A. Heuberger and J.M. Somers, Microcircuit Engineering 83 ed., H. Ahmed, J.R.A. Cleaver and G.A.C. Jones, Academic Press, London, 523 (1983).
[14] H. Betz, H.L. Huber, S. Pongratz, W. Rohrmoser, W. Windbracke and U. Mescheder, to be published.
[15] L. Csepregi, H. Seidel and H.J. Herzog, J. Electrochem. Soc., 131 2969 (1984).
[16] W.H. Brünger, H. Betz, A. Heuberger and J. Hersener, to be published.
[17] W. Windbracke, H. Betz, H.L. Huber, W. Pilz and S. Pongratz, to be published.
[18] K.H. Müller, P. Tischer and W. Windbracke, J. Vac. Sci. Technol. B4, 230 (1986).
[19] A. Heuberger, Z. Phys. B61, 473 (1985).
[20] A. Heuberger, Solid State Technol. 93, Feb 1986.
[21] W. Piltz, T. Sponholz, S. Pongratz and H. Mader, Microelectronic Engineering 3, 467 (1985).
[22] A.C. Tobey, Microelectr. Manufact. Testing, p.29, July 1986.
[23] A. Kluwe, K.H. Müller, H. Betz and H. Oertel, to be published.
[24] A. Wagner, SPIE-Conference on E-Beam, X-Ray and Ion-Beam Techniques for Submicron Lithographies, Ed. P. Blairs, 393 p.167, St. Clara, (1983).
[25] D.K. Atwood, G.J. Fisanick, W.A. Johnson and A. Wagner, SPIE-Conference on E-Beam, X-Ray and Ion-Beam Techniques for Submicron Lithographies, Ed. A. Wagner, 471, p.127, St. Clara, (1984).
[26] K. Gamo and S. Namba, Proc. Microcircuit Engineering '84, (1984), Berlin, Ed. Heuberger, Beneking, Academic Press p.389 (1985).
[27] P.J. Heard, J.R.A. Cleaver and H. Ahmed, J. Vac. Sci. Technol., B3,

87 (1985).
[28] D.J. Ehrlich, R.M. Osgood, P.J. Silversmith and T.F. Deutsch, IEEE Electron Dev. Lett. EDL-1, 101 (1980).
[29] H. Betz, A. Heuberger, N.P. Economou and D.C. Shaver, SPIE-Conference on Electron Beam, X-Ray and Ion Beam Techniques for Submicrometer Lithography, St. Clara, (1986).
[30] K.P. Müller, U. Weigmann and H. Burghause, to be published.
[31] J. Lorenz, J. Pelka, H. Ryssel, A. Sachs, A. Seidl and M. Svoboda, IEEE Trans. Computer Aided Design CAD-4, 421 (1985).
[32] S. Pongratz, H. Betz, A. Heuberger, Proc. Kodak Microelectronics Seminar, San Diego, p.143, (1983).
[33] H.L. Huber, H. Betz, A. Heuberger and S. Pongratz, Microcircuit Engineering '84, ed. A. Heuberger and H. Beneking, Academic Press, p.325 (1985)
[34] K.F. Dössel, H.L. Huber and H. Oertel, to be published.
[35] D.J. Nagel, Microelectronic Engineering 3, 557 (1985).
[36] H.C. Petzold and M. Kühne, Microelectronic Engineering 3, 565 (1986).
[37] J. Eberle, H. Krompholz, R. Lebert, W. Neff and R. Noll, Microelectronic Engineering 3, 611 (1985).
[38] J.S. Pearlman and J.C. Riordan, SPIE. 537 Electron-Beam, X-Ray and Ion-Beam Techniques for Submicrometer Lithographies IV, p.102, (1985).
[39] T. Kitayama et al., SPIE. 537, Electron-Beam, X-Ray and Ion- Beam Techniques for Submicrometer Lithographies V, (1986).
[40] A. Heuberger, Brookhaven Conf., SPIE Proc., 448, 8 (1983) ed. A.D. Wilson.
[41] H. Betz and G. Mülhaupt, Brookhaven Conf. SPIE Proc, 448, 83 (1983) ed. A.D. Wilson.
[42] M. Bieber, H.U. Scheunemann, H. Betz and A. Heuberger, J. Vac. Sci. Technol., 31, 1271 (1983).
[43] W.D. Klotz, K. Derikum, G.V. Egan Krieger, H.H. Flessner, H.G. Hoberg, H. Lehr, R. Maier, M. Martin, G. Mülhaupt, R. Richter, A. Schiele, L. Schulz, E. Weihreter and T. Westphal, Proc. of 11th Int. Conf. Cryogenic Engineering, Berlin 1986.
[44] Nikkei, Microdevice Magazine, p.67, April 1986.
[45] A. Wilson, Private Communication.
[46] E. Cullmann, Brookhaven Conf., Upton, N.Y., SPIE Proc., 448, 104 (1983).
[47] G. Doemens and P. Mengel, Siemens Forsch. Entwickl., Ber., 13, 43.
[48] K. Heinrich, H. Betz and A. Heuberger, J. Vac. Sci. Technol. B-1, 1352 (1983).
[49] H. Betz, A. Heinrich, A. Heuberger, H. Huber and H. Oertel, J. Vac. Sci. Technol. B-4, 248 (1986).
[50] L Csepregi, Microelectronic Engineering 3, 221 (1985).
[51] E.W. Becker, W. Ehrfeld, D. Münchmeyer, H. Betz, A. Heuberger, S. Pongratz, W. Glaashauer, H.J. Michel and R.V. Siemens: Naturwisschenschaften 69, 520 (1982).

Inst. Phys. Conf. Ser. No. 82
Paper presented at ESSDERC 1986, Cambridge 8–11 Sept. 1986

Silicon germanium MODFETs

E. Kasper, H. Daembkes

AEG Research Center, Sedanstr. 10, D-7900 Ulm, FRG

Abstract. In modulation doped SiGe/Si heterostructures, enhanced mobility is observed for electrons as well as for holes. The improved transport properties of the carriers are utilized for the fabrication of n-channel and p-channel modulation doped field-effect transistors. The importance of the built-in strain for the band ordering is emphasized. The fabrication and properties of the MODFETs are described.

1. Introduction

Development and availability of new crystal growth techniques such as the molecular beam epitaxy (MBE) give strong impulses to the creation of heterostructure materials. The low growth temperatures and small growth rate enable the precise control of layer composition and thickness down to even atomic scale. Completely new material systems can be grown and bandstructures can be tailored according to the specific applications.

The spatial separation of donor ions and free electrons by modulation doping techniques and the ability of growing specially designed potential barriers and wells have led to new transport properties in lateral as well as in vertical directions (Dingle et al. 1978, Tsuchiya et al. 1985). Devices based on these phenomena were already introduced like the high electron mobility transistor (HEMT, MODFET, TEGFET ...) (Mimura et al. 1980) and the resonant-tunnelling hot electron transistor (RHET) (Yokoyama et al. 1985). Many advantageous properties become possible in the area of optoelectronic devices.

The elimination of impurity scattering even at high carrier concentration levels has led to numerous constraints concerning field effect transistors from modulation doped heterostructures. Especially for analogue applications, excellent low noise and high frequency properties have been demonstrated with GaAs/AlGaAs HEMTs (Chao et al. 1985).

In parallel to the enormous MBE activity mainly on GaAs material some first successful results on MBE grown Si layers were reported (König et al. 1982). Meanwhile, Si MBE has reached production level and impressive results are achieved. In a feasibility study, large scale integrated GHz frequency dividers were produced from MBE grown layers (Kasper and Wörner 1984). The specific advantages of thin layer growth and abrupt doping transition are being exploited commercially for millimeter-wave IMPATT-diodes (Luy et al. 1986).

Because of its interesting potential for industrial application the heterosystem SiGe on Si substrate has gained increasing attention. Like the III-V heterostructure this Si based heterosystem promises advantageous properties for electronic devices as well as for optoelectronic purposes. Depending upon the application, either the improved transport properties or the optical properties of the alloy will be utilized. Improved room temperature transport properties and experimental evidence of zone folding are already demonstrated (Jorke and Herzog 1985, Brugger et al. 1986). Experimental modulation-doped field effect transistors (Daembkes et al. 1985, 1986, Pearsall and Bean 1986) and optoelectronic detectors (Pearsall et al. 1986) could already demonstrate device level performance. The ultimate goal will be the tailoring of material properties for device optimization by superposition of dopant and heterostructures and by Brillouin zone folding caused by the superlattice periodicity. This might finally lead to the predicted direct band gap of SiGe superlattices (Gnutzman and Clausecker 1974).

A fundamental difference of the SiGe material from the well-known III-V heterosystems is its large lattice mismatch to the Si substrate. At the same time, this mismatch and the resulting strain in the layers are one of the keys to its interesting physical properties. In this article it will be shown that the ability to grow strained layers with precisely controlled strain in the heterostructure is of great importance for the fabrication of modulation doped field effect transistors in the SiGe/Si heterosystem.

2. Modulation doping

The mobility of carriers in uniformly doped bulk material is limited by scattering from phonons and from ionized impurities. The upper limit of the mobility of the undoped semiconductor material is given by the phonon

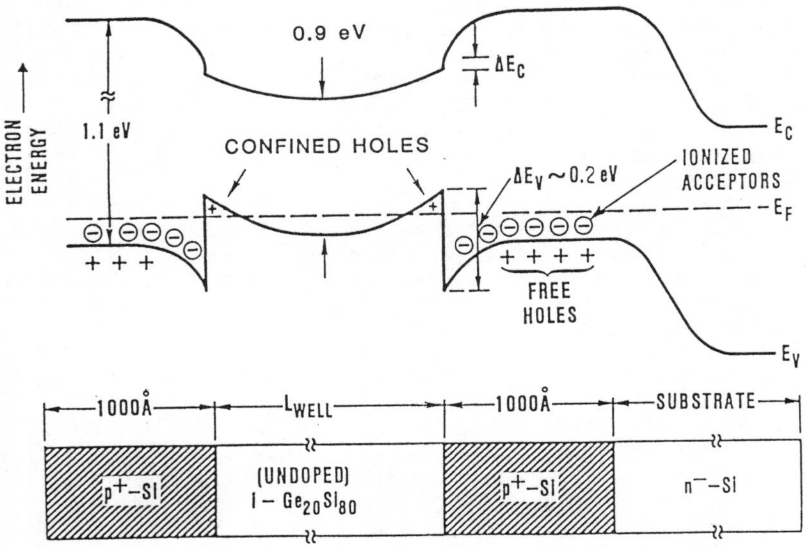

Fig. 1: Scheme and band edges of a modulation doped p-Si/SiGe (by courtesy of J.C. Bean).

scattering (thermal limitation). At room temperature this upper limit of bulk mobility in silicon amounts to about 1500 cm^2/Vs and 450 cm^2/Vs for electrons and holes, respectively. By doping, the mobility in silicon doped to a concentration of $10^{18}/cm^3$ decreases considerably, e.g. to about 275 cm^2/Vs and 140 cm^2/Vs for electrons and holes.

A promising way to overcome the limitations arising from ionized impurity scattering consists of replacing uniformly doped material by modulation doped heterostructures.

2.1 Principle of modulation doping

Consider a structure as given in Fig. 1.

Conduction band edge E_C, valence band edge E_V and Fermi energy level E_F are shown as functions of the position across the modulation doped heterostructure. Band edge discontinuities match the different band gaps on both sides of the interface (ΔE_v and ΔE_c for valence band and conduction band, resp.)

The structure on top of a silicon substrate consists of a series of doped silicon/undoped SiGe/doped silicon .... In this case the lower bandgap material SiGe is undoped. It is assumed that some of the holes generated by the acceptors in the wider gap material Si jump across the heterointerface. These confined holes create a 2-dimensional carrier gas at the undoped side of the heterointerface. Spatial separation of scattering centers (ionized acceptors) and carriers (confined holes) results in mobility enhancement compared to uniformly doped bulk material.

2.2 2-dimensional hole gas

People et al. (1984) started the investigation of modulation doping in SiGe/Si heterostructures which were grown by molecular beam epitaxy in a configuration as shown in Fig. 1. The electrical results demonstrated clearly the existence of a 2-dimensional hole gas but they failed to get a 2-dimensional electron gas. These experimental findings were explained by a near flat conduction band ($\Delta E_c \ll \Delta E_v$) and a large valence band discontinuity ΔE_V. The existence of the hole gas influences the sheet concentration and mobility of the carriers as measured by conductivity and Hall effect. The sheet concentration of holes in the modulation doped p-Si/SiGe decreases with decreasing temperature because of freeze out of the free carriers and saturates then on the level of the confined carriers of the 2-dimensional (2 D)-hole gas (Fig. 2).

The range of the Coulomb-field of ionized impurities is rather large. It is therefore advantageous to design the doped side of the heterojunction with doping setback of width w. Generally, the 2 D-carrier gas concentration decreases, but the mobility increases by using a doping setback. The undoped region is also called a spacer. . The influence of doping setback on mobility is shown in Fig. 3.

A direct proof of the existence of a two-dimensional carrier gas is obtained by magneto-resistance measurements (Shubnikov-de Haas oscillations for a magnetic field H perpendicular to the layer, Fig. 4).

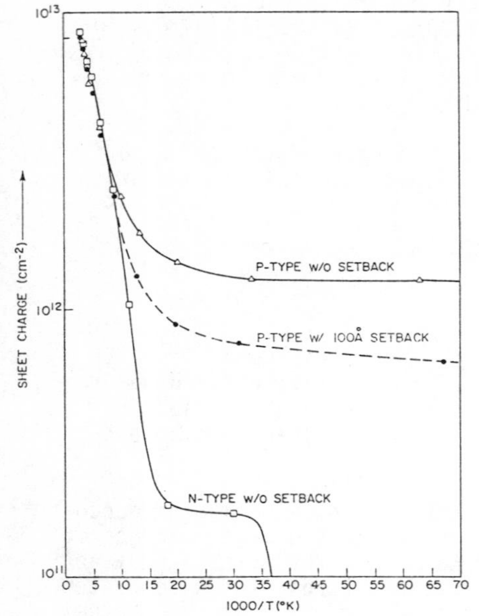

Fig. 2: Sheet concentration versus temperature for the modulation doped structure in Fig. 1 (by courtesy of J.C. Bean). The 2 D-hole gas does not freeze out. For influence of the dopant setback see text. There is no significant indication of a 2 D-electron gas.

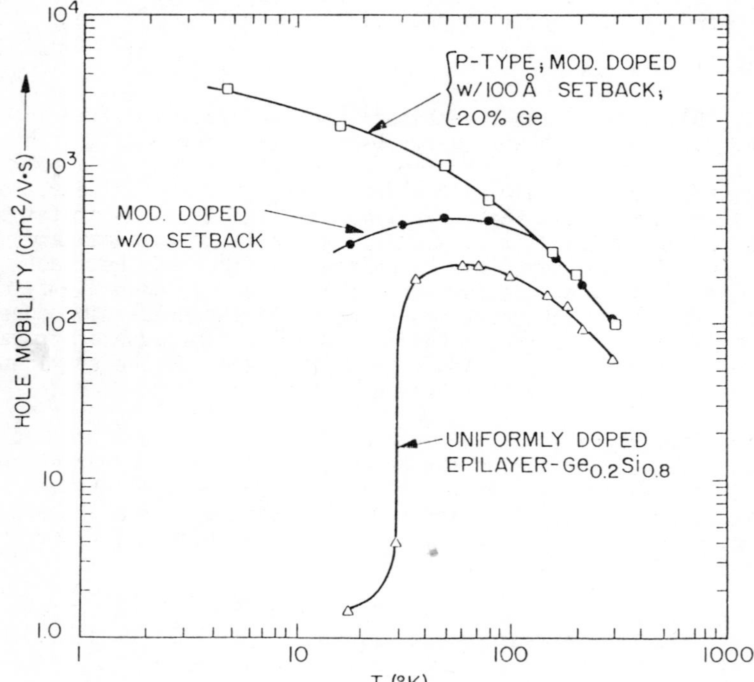

Fig. 3: Hole mobility μ versus temperature T for modulation doped p-Si/SiGe (by courtesy of J.C. Bean).

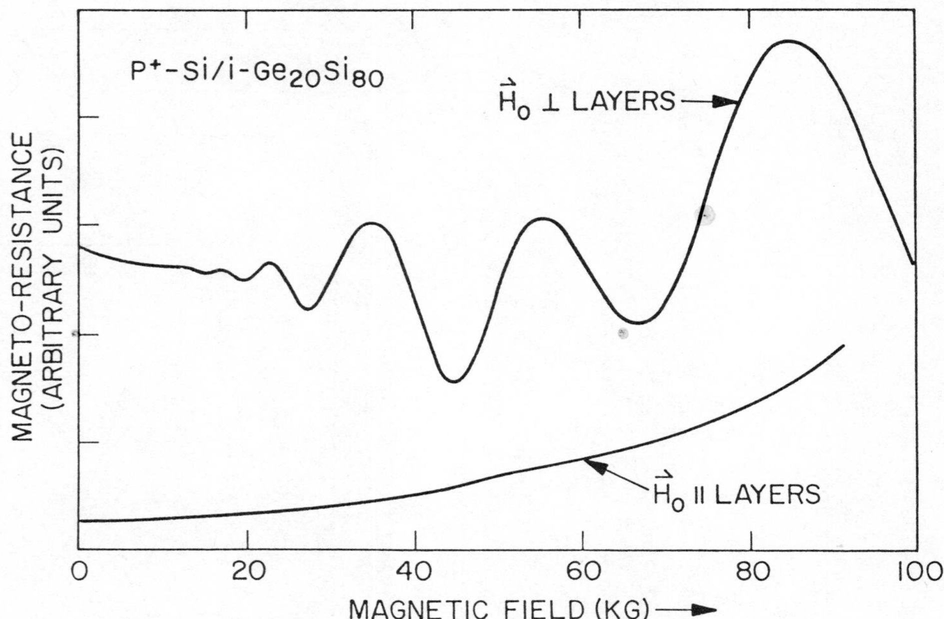

Fig. 4: Magneto-resistance of modulation doped p-Si/SiGe (measured at T = 1.8 K) (by courtesy of J.C. Bean). Shubnikov-de Haas oscillation for hot layers indicate a 2-dimensional hole gas with a sheet concentration of about $3.5 \times 10^{11}/cm^2$.

2.3 2- dimensional electron gas

In order to grow Si/SiGe heterostructures with symmetrical strain our group performed MBE of heterostructures or superlattices on thin incommensurate SiGe buffer layers on top of the silicon substrates (Eichinger et al. 1985, Kasper et al. 1986 , Ricker and Kasper 1986). In such heterostructures we found strong evidence for the existence of a two-dimensional electron gas (Jorke, Herzog 1985, Abstreiter et al. 1985) in apparent contradiction to the above results of the Bell group. The evidence for a 2- dimensional electron gas was given by the temperature dependence of the sheet concentration, by mobility enhancement compared to doped bulk material (Fig. 5) and by magneto-resistance measurement showing clear Shubnikov-de Haas oscillations for magnetic field perpendicular to the layers.

Technically important seemed to be the room temperature enhancement of the mobility (e.g. a factor of four for silicon with a mean doping density of $4 \times 10^{18}/cm^3$) and the large sheet concentration of $3 \times 10^{12}/cm^2$ of confined electrons. The influence of the position of the n-type dopant Sb was investigated systematically by varying the position of a doping spike (Jorke, Herzog 1985) according to Fig. 5b. Surprisingly, the 2-dimensional electron gas was generated in silicon by doping the low band gap material SiGe.

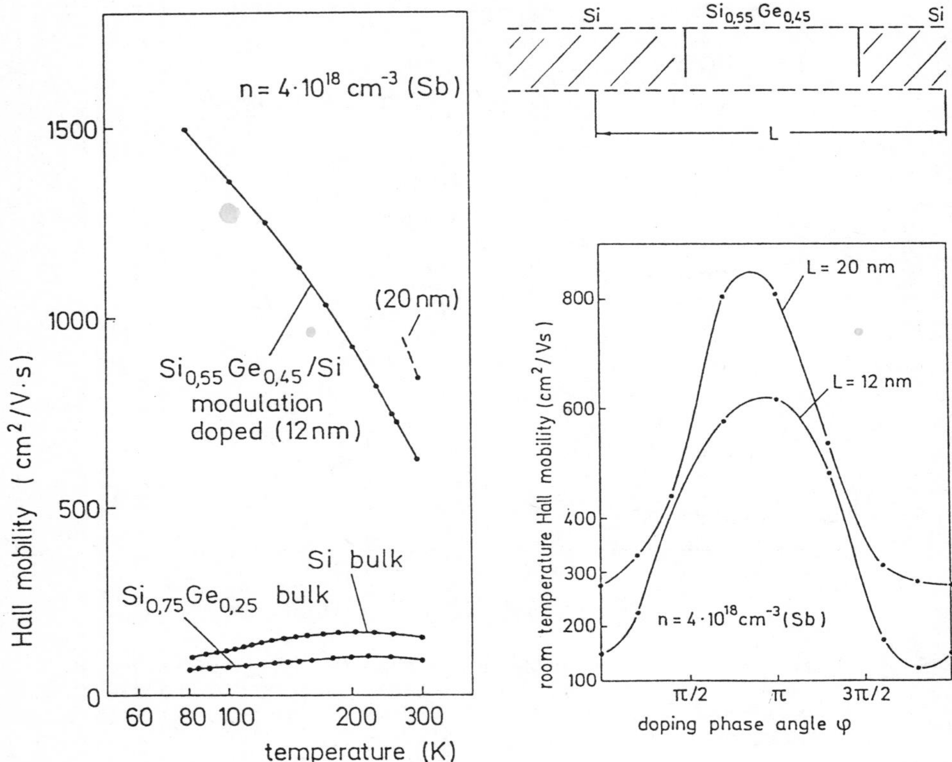

Fig. 5: Electron mobility (measured by Hall-effect) of modulation doped n-Si/SiGe as function of temperature (left) and doping phase angle (right), respectively. Clearly shown are the enhanced mobility even at room temperature and the strong dependence on position of the doping spike.

3. Strain adjustment as key to 2 D-electron gas in modulation doped Si/SiGe

The AEG group and the Bell group made very different observations concerning the existence of a 2 D-electron gas in modulation doped Si/SiGe structures. The only essential experimental difference was the strain situation in the different structures (asymmetrical strain in the SiGe layers for the Bell experiments, symmetrical strain for the AEG experiments). Consequently, the role of strain in the Si/SiGe heterosystem was investigated more thoroughly.

3.1 Mismatch accommodation in strained layer systems

Both, Si and Ge atomic lattices are of the diamond type with lattice constants of 0.543 nm and 0.565 nm, respectively. The mismatch between the two lattice cells amounts to 4.2 %. The SiGe alloy obeys Vegard's law which predicts a linear relationship between Ge content and lattice constant. Nature knows two answers to accommodation of mismatched heterosystems. The elasticity answer is accommodation by strain. The plasticity answer is accommodation by misfit dislocations. Fig. 6 compares these two ways to accommodate mismatch.

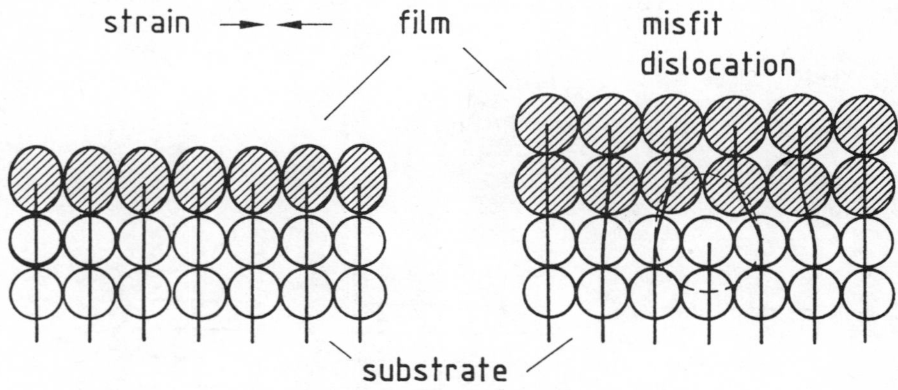

Fig. 6: Accommodation of a mismatched heterosystem. For simplicity a cubic lattice is assumed. Left side: strain. Right side: misfit dislocations.

For very thin layers mismatch accommodation by strain occurs (Kasper et al. 1975). For thicker layers accommodation by misfit dislocations gains increasing importance. This was explained by v.d. Merwe (1972) on the basis of a thermodynamic equilibrium theory. A critical thickness was predicted, below which accommodation is purely by strain, and above which misfit dis-

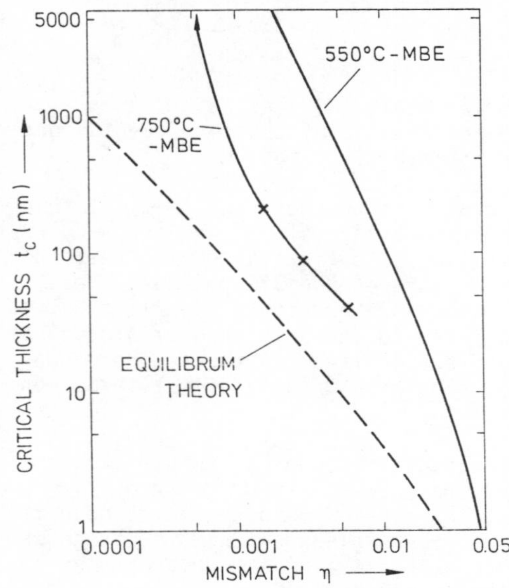

Fig. 7: Critical thicknesses versus lattice mismatch. Compared are equilibrium theory with experimental data for MBE at 750°C and 550°C, respectively.

locations are generated releasing the strain. The experimentally found critical thicknesses are higher than those predicted by equilibrium theories (Kasper and Herzog, 1977). With MBE at temperatures as low as 550° C the critical thickness can be further increased (Bean at al. 1983). This allows one to grow a strained layer system with a rather large lattice mismatch if the thickness of a strained layer is below the critical thickness (Fig. 7).

3.2 Strain adjustment

Consider a structure as shown in Fig. 8

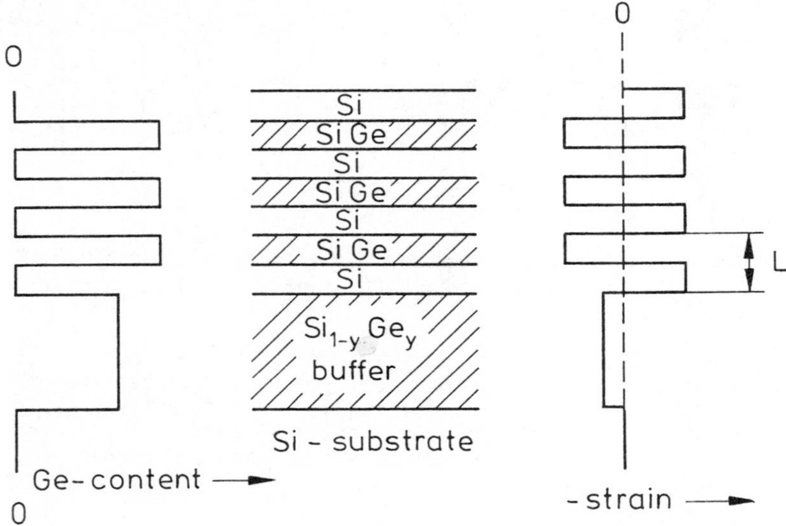

Fig. 8: Strain symmetrization: Superlattice $Si/Si_{1-x}Ge_x/Si$... on a buffer layer $Si_{1-y}Ge_y$ on top of a silicon substrate. Left side: Ge-distribution. Right side: Strain distribution.

The thickness of a buffer layer is chosen to be above the critical thickness. Part of the mismatch is accommodated by a misfit dislocation network lying at the interface between substrate and buffer layer. The residual strain changes the natural lattice constant of the SiGe buffer. The strained lattice constant of the incommensurate buffer layer lies between the natural lattice constants of silicon and the silicon germanium alloy with Ge content y. The incommensurate buffer layer on the silicon substrate can be considered as an alloy substrate with an effective Ge content y*.

For calculation of the effective Ge content y* one has to use experimentally determined values of the film strain as a function of layer thickness and growth temperature, because existing theories do not predict quantitatively correct values of film strain. Fig. 9 gives an example of the calculation for 550° C growth temperature using strain data measured by Rutherford back scattering (Bean et al. 1983, 1987).

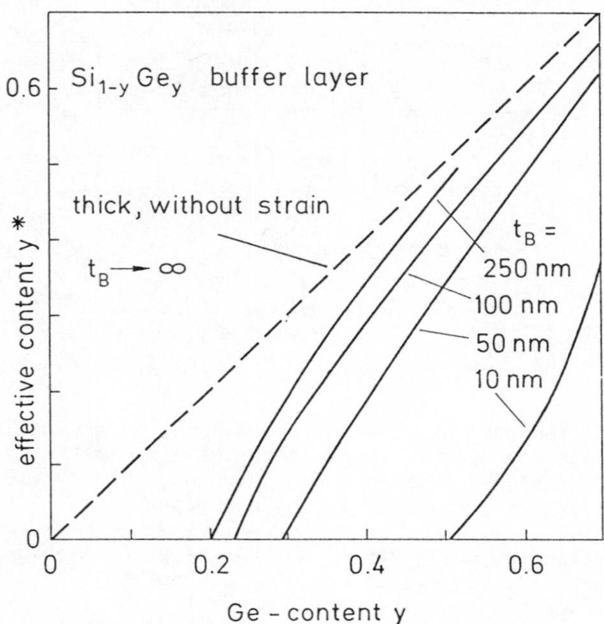

Fig. 9: The strained, incommensurate SiGe buffer layer (Ge content y) on top of the silicon substrate provides an in-plane lattice constant which equals the lattice constant of an unstrained SiGe alloy with an effective Ge content y*. The effective Ge content y* is calculated as function of the Ge content y of the buffer layer for various buffer layer thicknesses t_B.

The heterostructure or superlattice on the buffer layer is strained to match the alloy lattice constant given by the effective Ge content y*. By varying the parameters of the buffer layer (Ge content, thickness, growth temperature) one can adjust the strain distribution in the heterosystem.

Especially important is a symmetrical strain distribution with tensile strain in the Si-layers and compressive strain of the same magnitude in the SiGe layers. This would lead to a stable configuration with minimum strain energy.

3.3 Electronic band ordering

A rough idea about ordering of the electronic bands at a heterointerface is provided by the electronegativity of the materials. For the system Si/SiGe one would expect a strong discontinuity ΔE_v at the valence band, but nearly a flat conduction band ($\Delta E_c \ll \Delta E_v$) as was confirmed by the Bell experiments. From our experiments one would expect a staggered band ordering as shown in Fig. 10.

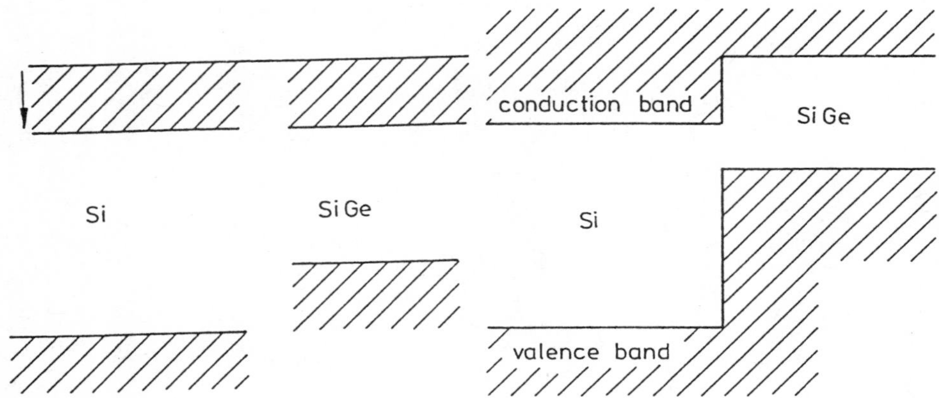

Fig. 10: Scheme of the band ordering at the Si/SiGe interface. Left side: Flat conduction band as suggested by electronegativity arguments. Right side: Staggered band ordering which would result in a 2 D-electron gas in Si as observed by our group.

The key for understanding was given by Abstreiter et al. (1985, 1986). They argued that the sixfold degenerate conduction band is split into two-fold and four-fold degenerate valleys by the in-plane strain. They stated that symmetrization of the strain would change the type of band ordering from flat conduction band to staggered band ordering. This was confirmed by calculations made at the Bell group (People and Bean 1986). It is now clear that results of both groups were in apparent discrepancy because of the different strain distribution in the experiments. A two-dimensional electron gas in silicon needs a staggered band ordering obtained by strain adjustment.

4. Fabrication of MODFETs

4.1 P-channel MODFETs

P-channel $Si_{1-x}Ge_x$/Si modulation-doped FETs were for the first time demonstrated by Bell researchers (Pearsall et al. 1985). According to their results concerning the confinement of holes in $Si_{1-x}Ge_x$ (People et al. 1984) they used the following layer structure: on a p -Si substrate an undoped buffer layer was grown followed by an undoped 25 nm $Ge_{0.2}Si_{0.8}$ confinement layer. The doping is incorporated in the following 50 nm thick p^{-}-Si layer as a modulation doped layer of some ten nm thickness which is typically set back from the heterointerface by about 10 nm. The thin doping layer is formed during the MBE growth by a simultaneous implantation of boron into the undoped Si layer (Bean and Sadowski 1982). Typical implantation energies are around 1 keV and the dose gives bulk concentrations around 2 x 10^{18} cm^{-3} (Pearsall et al. 1985). The typical MBE structure and the proposed band structure are shown in Fig. 11. As can be seen two potential wells are formed at the heterointerfaces, but it is assumed that most of the carriers are confined to the side of the well near the boron doping. Only near a pinch-off condition will a significant current flow through the lower channel. This fact might lead to a slightly degraded pinch-off behaviour.

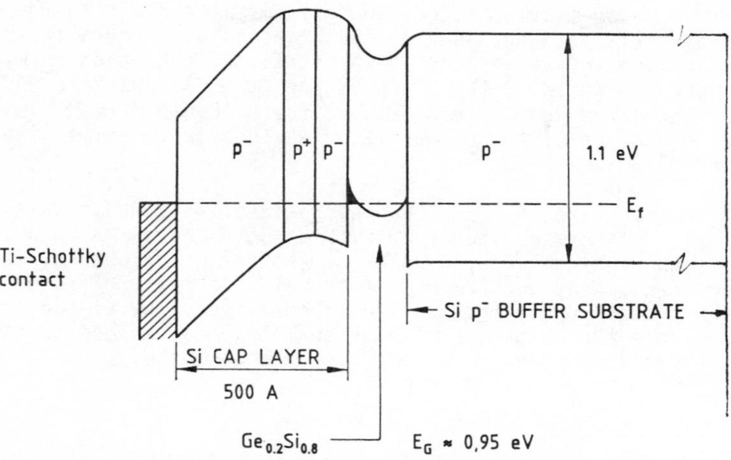

Fig. 11: Proposed schematic band diagram for the $Si_{1-x}Ge_x$ p-channel MODFET. Two two-dimensional hole gases are formed.

The FETs are fabricated using standard Si VLSI processing techniques. Ohmic contact regions were prepared using BF_2 implantation. The annealing temperature was kept at modest levels (700°C - 750°C) in order to limit the defect formation due to strain relaxation and to minimize a possible Ge diffusion. As shown in Fig. 12, Al is used for ohmic contacts and Ti to form the Schottky gate. Layers were thinned using reactive-ion etching techniques (Pearsall and Bean 1986).

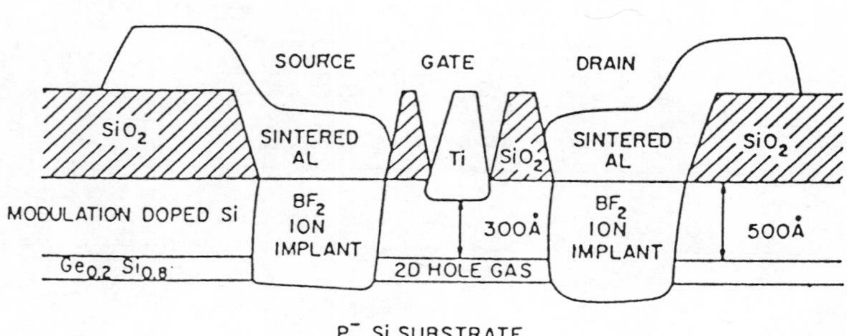

Fig. 12: Cross-sectional sketch of the processed $Si_{1-x}Ge_x/Si$ p-channel MODFET (courtesy J.C. Bean).

4.2 N-channel MODFETs

From our experiences with the enhanced electron mobility in the multiquantum well structures (Abstreiter et al. 1985) we developed the following simple structure for an n-channel modulation-doped SiGe/Si FET as shown in Fig. 13. On a (100) high resistivity Si substrate a 200 nm $Si_{0.7}Ge_{0.3}$ buffer layer is grown. Its thickness is well above the critical value for stress accommodation by misfit dislocations. It is intended that the strain at this heterointerface is released by the creation of interfacial misfit dislocations. By TEM investigations it was confirmed that a dense misfit dislocation network exists at the plane between the Si substrate and the buffer layer. The subsequent layers, however, are virtually free of these defects. The Ge content of the buffer layer was chosen to symmetrize the strain in the following layers.

On top of the buffer layer a 20 nm undoped Si layer follows in which the quantum well will be formed. Then a 10 nm $Si_{0.5}Ge_{0.5}$ layer is grown. This layer contains the antimony doping spike of about 2 nm thickness. To avoid the formation of a second quantum well a graded SiGe layer of 10 nm thickness forms the transition to the undoped Si top layer. The Si top layer was grown to enable the formation of high-quality Schottky contacts for the gate electrode and to avoid unwanted surface currents.

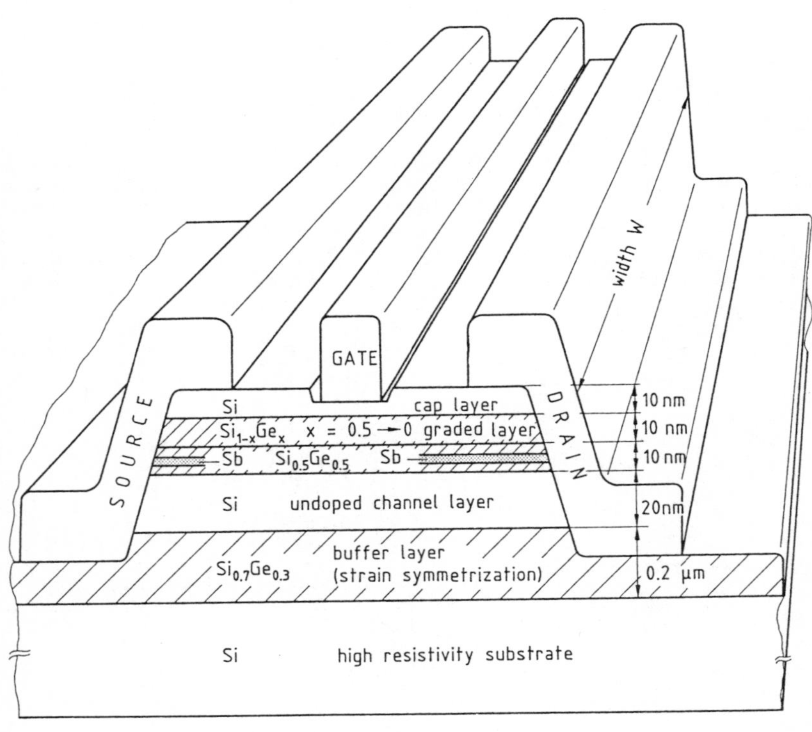

Fig. 13: Cross sectional sketch of the typical n-channel SiGe MODFET structure.

The layers were grown in our standard SiGe MBE equipment at a substrate temperature of 600°C and a growth rate of about 0.14 µm/h on high resistivity Si substrates. All the layers except for the delta-doped SiGe layer are undoped. The Sb doping spike was produced using the method of doping by secondary implantation (DSI) (Jorke et al. 1985) during the growth of the $Si_{0.5}Ge_{0.5}$ layer.

The proposed band structure of this n-channel structure is depicted in Fig. 14. We assume that only one conductive channel is formed at the $Si_{0.5}Ge_{0.5}/Si$ heterointerface due to the band bending of the buffer layer.

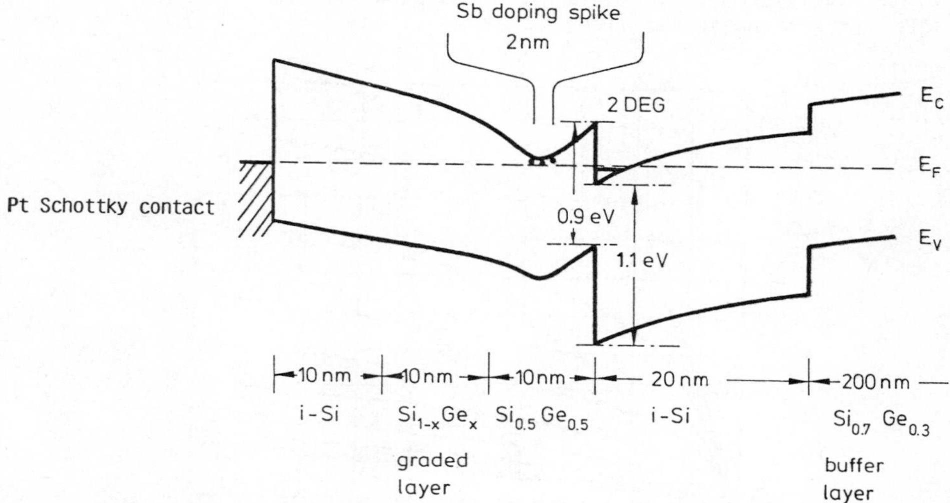

Fig. 14: Band-structure as proposed for the n-channel SiGe/Si MODFET.

Further technological steps were made in analogy to well proven III-V MODFET technology (Daembkes et al. 1984). Special care was taken to avoid any high temperature treatment which might degrade the abruptness of the heterointerface or promote diffusion. Conventional contact lithography was used to define the device levels. The individual devices were isolated by a mesa technique. A dry etching process using CF_4 + 3.9 % O_2 in a barrel type reactor was applied. Ohmic contacts were formed by thermal evaporation of 0.3 µm AuSb and lifting off the excess material. Due to the undoped Si top layer and the omission of any n^+-implantation or diffusion, the ohmic metallization has to be alloyed. Alloying at 330°C under protective gas for 30 seconds leads to low resistivity ohmic contacts with smooth edges and nearly perfect morphology. The gate was formed by e-gun evaporation of a Pt/Ti/Au sandwich of 100 nm/100 nm/ 150 nm thickness. Platinum is chosen as a Schottky metal as it establishes one of the highest Schottky barriers to Si. Immediately before the evaporation an oxygen plasma is used to remove organic residuals. Then a slight recessing is followed by an HF dip to remove oxides.

5. Electrical characterization

5.1 P-channel devices

By adjusting the layer thickness under the gate contact both enhancement-mode and depletion mode p-channel MODFETs were fabricated (Pearsall and Bean 1986). A typical characteristic for a depletion mode transistor is shown in Fig. 15. Maximum values of the transconductance are reported to be slightly in excess of 3 mS/mm for gate lengths around 2 µm. The achievable values are reduced by the fairly high parasitic lead resistances of the FET structure used. The reasons for this are the low mobility of the holes, the low sheet carrier concentration of about 2.5 x 10^{11} cm^{-2}, and the long source to drain spacing. Further inspection of the output characteristics reveals a certain amount of output conductance which is disadvantageous for device applications. Also the pinch-off behaviour is not yet satisfactory. The origin of this parallel conducting channel is not yet completely clear.

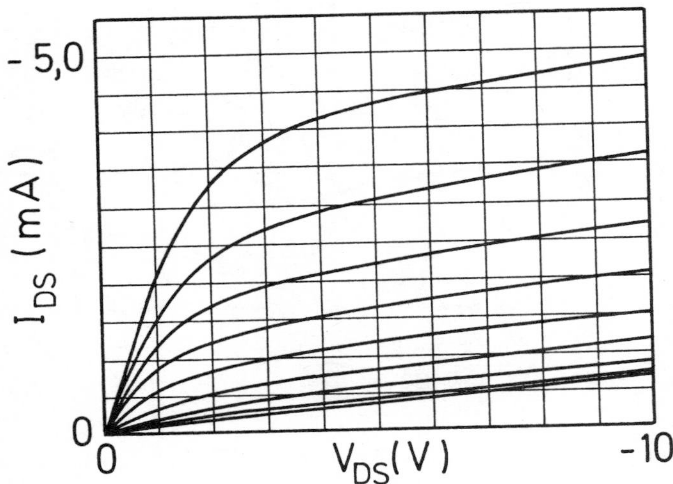

Fig. 15: Electrical characteristics of an enhancement-mode p-channel SiGe/Si MODFET. ΔV_{GS} = 1 V

The hole transport properties were extracted from the dc characteristics of the transistors. At zero gate bias the drift velocity of the holes in the channel was determined to be v_D = 1.1 x 10^7 cm/s which is in good agreement with expected values (Pearsall and Bean 1986).

5.2 N-channel devices

For the fabrication of n-channel MODFETs a number of different wafers with fairly low carrier concentrations in the range of 1 - 6 x 10^{11} cm^{-2} were grown. All of the investigated samples resulted in good FETs. Typical dc-characteristics of a SiGe/Si n-channel MODFET are shown in Fig. 16.

All devices showed good characteristics with a complete pinch-off behaviour, a distinct ohmic and saturated region. No looping was detected. In some cases a certain bias-dependent shift of the characteristics towards higher currents was observed which might indicate the presence of some traps.

The best values of the extrinsic transconductance were found to be about 50 mS/mm for 2 µm gate length. Analysis of the device characteristics and a number of special test arrangements reveals that the measured device behaviour is dominated by high parasitic resistances. They drastically reduce the performance of the extrinsic device. This limitation is due to our first approach of omitting any optimization of the device layout and processing such as n^+-contact layers or n^+-implantation, which, if not well tested, might degrade the abruptness of the heterointerface.

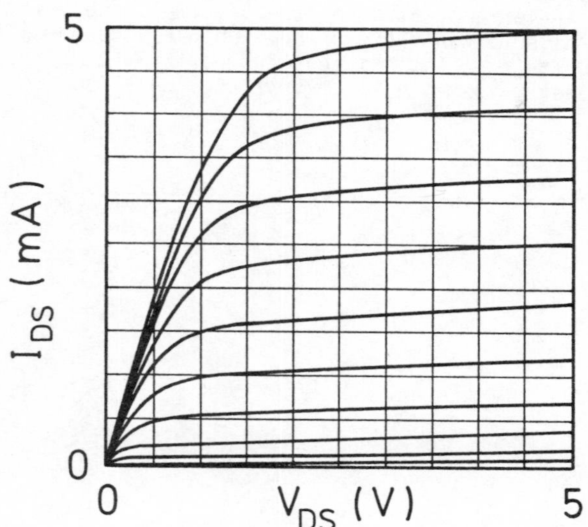

Fig. 16: Typical characteristics of an n-channel SiGe/Si MODFET.
L = 2 µm W = 90 µm
Upper trace corresponds to V_{GS} = 0V, ΔV_{GS} = 0.2 V

Further technological runs using optimized layout and processing will clearly improve the extrinsic performance. From the gate length dependence, as obtained from our test structures and from the dependence upon the carrier concentration of the different samples, we expect that a maximum transconductance of 180 mS/mm is achievable for a sheet carrier concentration of 1 - 2 x 10^{12} cm^{-2} and a gate length of 1 µm.

Regarding the layer sequence and our non-optimized device fabrication, the measured good device performance is assumed to be due to the improved transport properties of the electrons in a quantum well channel. The shape of the conductive channel was tested by conventional C-V carrier profiling at room temperature. Gate diodes of the FETs as well as specially designed test diodes were used and led to the results as shown in Fig. 17 and Fig. 18.

No Debye correction was applied. A very sharp carrier profile was measured, though the shape obtained should be interpreted with some reservation as the influence of a possibly emerging gate diffusion capacitance is not yet subtracted. The peak of the carrier concentration is observed exactly at the position of the grown heterointerface. Depending upon the incorporated amount of doping material in the different samples, peak carrier concentrations were measured in the range of 1 x 10^{18} cm^{-3}. The profile indicates that all the carriers are confined in a very narrow part of the undoped Si layer. This can only be explained by the existence of a very sharp potential well.

The transport properties of the electrons in the finished MODFETs were investigated by means of the magnetotransconductance measurement (Prost et al 1986). Very high values of the mobility ranging from 1200 cm^2/Vs to more than 1500 cm^2/Vs were measured at room temperature for devices with more than 10^{18} cm^{-3} carrier concentration (peak value). These values are a factor of 6 - 8 higher than those of homogeneously doped silicon material. This leads to the expectation, that the performance of the SiGe/Si MODFETs will be at least comparable to conventional Si FETs.

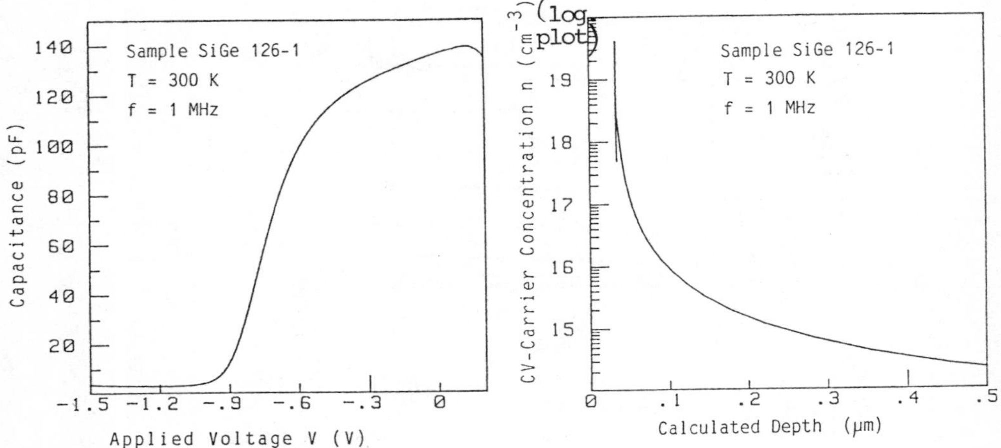

Fig. 17: Measured capacitance of a test diode versus the applied voltage.

Fig. 18: Logarithm of the carrier concentration versus the depth as obtained by CV measurement. The peak is exactly at the position of the grown heterointerface.

To test the frequency behaviour to the new devices, preliminary high frequency measurements were carried out. From the measured scattering parameters a maximum frequency of 3 GHz was calculated for devices with 2 µm gate length. This indicates that the conductive channel of our structures is maintained up to microwave frequencies. A small signal equivalent circuit was calculated from the measured data (Fig. 19). It clearly reveals the presently limiting elements and gives valuable hints for the further device optimization. Beside the reduction of the gate length, which leads to a reduced input capacitance and a higher transconductance, the reduction of the parasitic resistances has to be the main objective. Our experiences with the III-V MODFET show, however, that even excellent 2DEGs alone will not lead to the desired low sheet resistances and additional means have to be used.

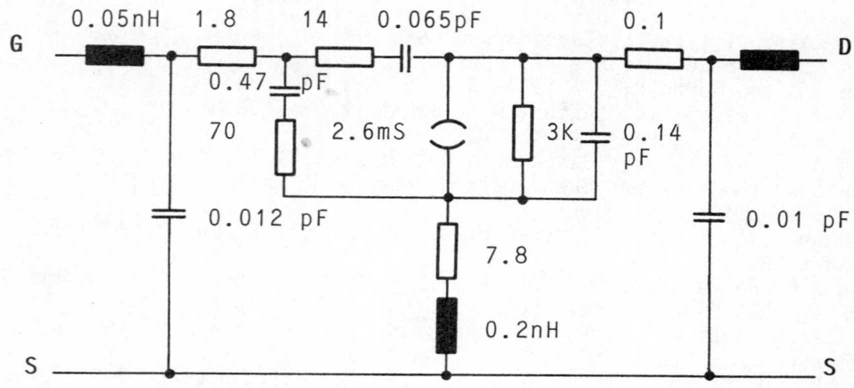

Fig. 19: High frequency equivalent circuit of an n-channel SiGe/Si MODFET. L = 2.1 µm W = 290 µm V_{GS} = - 0.8V

6. Conclusion

After the phase of pure growth and investigation of the new $Si_{1-x}Ge_x/Si$ heterosystem the time of first applications has come. The interesting transport properties are successfully transferred into working devices. The results obtained by the first attempts give rise to expected improvements of the present day FET performance. A combination of p- and n-channel FETs suggests complementary circuits e.g. CMOS, which might be faster than todays fastest circuits. On the way to their realization the existing technology and device concepts have to be made consistent, possibly by adaptation of the GaAs/AlGaAs approach (Cirillo et al. 1985).

A second very important field of application will be the combination of optoelectronic elements such as detectors and waveguides with fast electronic control circuits. All these ingredients are already demonstrated separately. This development might lead to the first real mass production of optoelectronic integrated circuits in the 1.3-1.5 µm wavelength region since one could take advantage of the full power of the mature silicon technology.

To fully exploit the silicon experience the device concepts should be revised. For the first tests the Schottky-gate approach led to an early success. But as the surface of the MODFET structures is silicon again, one should avoid the difficulties with contact formation and low barrier heights and take full advantage of the natural oxide properties. The advances in low temperature oxide formation and the bandstructures and experimental results of the SiGe/Si heterostructure demonstrated in this article, suggest that this will be the next step to fully silicon technology compatible devices.

Acknowledgement

The authors express their thanks to H. Kibbel, H. Jorke, H-J. Herzog and H. Hieber for their valuable contributions. J.C. Bean is acknowledged for the information about p-channel MODFETs.

Abstreiter G, Brugger H, Wolf T, Jorke H and Herzog H J 1985, Phys. Rev. Lett. 54 2441

Abstreiter G, Brugger H, Wolf T, Zachai R and Zeller Ch 1986, Two-Dimensional Systems: Physics and New Devices (Berlin: Springer Series in Solid-State Sciences 67) p. 130

Bean J C and Sadowski E A 1982, J. Vac. Sci. Technol. 20 137

Bean J C, Feldmann L C, Fiory A T, Nakahara S and Robinson I K 1984, J. Vac. Sci. Technol. A2 436

Bean J C 1987, Silicon Based Heterostructures in Silicon Molecular Beam Epitaxy (Boca Raton; CRC), Ed.: Kasper E and Bean J C, will be published

Brugger H, Abstreiter G, Jorke H, Herzog H J and Kasper E 1986, Phys. Rev. B33 5929

Chao P C, Palmateer S C, Smith P M, Mishra U K, Duh K H G and Hwang J C M 1985, IEEE Electron Dev. Lett. EDL-6 No 10 531

Cirillo N, Shur M S, Vold P J, Abrokwah J K and Tufte O N 1985, IEEE Electron Dev. Lett. EDL-6 No 12 645

Daembkes H, Brockerhoff W, Heime K, Ploog K, Weimann G and Schlapp W 1984 El. Lett. 20 No 15 615

Daembkes H, Herzog H J, Jorke H, Kibbel H and Kasper E 1985, IEDM Technical Digest 768

Daembkes H, Herzog H J, Jorke H, Kibbel H and Kasper E 1986, IEEE Trans. Electron Dev. ED-33 No 5 633

Dingle R, Störmer H L, Gossard A C and Wiegmann W 1978, Appl. Phys. Lett. 33 665

Eichinger P, Frenzel E, Iberl F, Kasper E and Kibbel H 1984, Proc. 1st Int. Symp. Si-MBE, Vol. 85-7 367, Electrochem. Soc. , Ed.: Bean J C

Gnutzman H and Clausecker K 1974, Appl. Phy. 3 9

Jorke H and Herzog H J 1985, Proc. 1st Int. Symp. Si-MBE (Toronto) Vol. 85-7, 352 Electrochem. Soc. , Ed.: Bean J C

Jorke H, Herzog H J and Kibbel H 1985, Appl. Phys. Lett 47 511

Kasper E, Herzog H J and Kibbel H 1975, Appl. Phys. 8 199

Kasper E and Herzog H J 1977, Thin Solid Films 44 357

Kasper E and Wörner K 1984, Proc. 2nd. Int. Symp. VLSI Sci. Technol. 1984, Vol. 84-7 429, Electrochem. Soc. , Ed.: Bean K E and Rozgonyi G A

Kasper E, Herzog H J, Dämbkes H and Ricker Th 1986, Two-Dimensional Systems: Physics and New Devices (Berlin: Springer Series in Solid-State Sciences 67) 52

König U, Herzog H J, Jorke H, Kasper E and Kibbel H 1982, Collected Papers of MBE-CST-2, 193 (Tokyo: Jap. Soc. Appl. Phys.)

Luy J F, Kibbel H and Kasper E 1986, Int. J. Infrared and Millimeter Waves 7 305

Mimura T, Hiyamizu S, Fujii T and Nanbu K 1980, Jap. Journ. Appl. Phys. 19 No 5 L225

Pearsall T P, Bean J C, People R and Fiory AT 1985, Proc. 1st Int. Symp. Si-MBE (Toronto) Vol. 85-7 400, Electrochem. Soc. , Ed.: Bean J C

Pearsall T P and Bean J C 1986, IEEE Electron Dev. Lett. EDL-7 No 5 308

Pearsall T P, Temkin H, Bean J C and Luryi S 1986, IEEE Electron Dev. Lett. EDL-7 No 5 330

People R, Bean J C, Lang D V, Sargent A M, Störmer H L, Wecht K W, Lynch R L and Baldwin 1984, Appl. Phys. Lett 45 1231

People R and Bean J C 1986, Appl. Phys. Lett. 48 538

Prost W, Brockerhoff W, Heime K, Ploog K, Schlapp W, Weimann G and Morkoc H 1986, IEEE Trans. Electron Dev. ED-33 No 5 646

Ricker Th and Kasper E 1986, Semiconductor Quantum Well Structures and Superlattices (Les Ulis Cedex, France: Led iditions de physique) Ed.: Ploog K and Linh N T 193

Tsuchiya M, Sakaki H and Yoshino J 1985, Jap. Journ. Appl. Phys. 24 L466

Yokoyama N, Imamura K, Muto S, Hiyamizu S and Nishi H 1985, Jap. Journ. Appl. Phys. 24 No 11 L853

Inst. Phys. Conf. Ser. No. 82
Paper presented at ESSDERC 1986, Cambridge 8–11 Sept. 1986

Device aspects of megabit RAMs

F.M. Klaassen
Philips Research Laboratories
Eindhoven, The Netherlands

ABSTRACT

Device aspects of submicron technology for megabit RAMs are discussed from the viewpoint of applied physics. Successively considered are the n-channel MOSFET, the p-channel MOSFET, the dynamic RAM cell, the static RAM cell, the device isolation and the interconnection.

INTRODUCTION

The ability to increase integration density, improve circuit performance and reduce costs made possible by scaling of active devices and process innovations has been the economic and technical driving force of MOS technology. Whereas the decade of the seventies was dominated by n-MOS technology, chip and system power considerations are now favouring CMOS as the leading technology for VLSI. Since for a practical realization of megabit level complexity device dimensions in the submicron range are required, several specific problems arise:

- control of hot carrier effects and punch through in active devices,
- decrease of signal/noise ratio and soft error immunity of memory cells,
- effects of less ideal scaling of the isolation and interconnection.

In this paper device aspects of submicron technology for megabit RAMs are discussed from the viewpoint of applied physics. Successively we consider the n-channel MOSFET, the p-channel MOSFET, the dynamic RAM cell, the static RAM cell, the isolation and the interconnection of devices.

N-CHANNEL MOSFET

When the structural dimensions of n-channel MOSFETs are reduced to the submicron level, two conflicting requirements have to be met. First, the gate drain/source charge-sharing effect (1,2) and the associated drain-induced potential barrier lowering (3) [which may cause a decrease of threshold voltage and an increase of current in the off-state], require the gate insulator thickness and junction depth to be reduced. On the other hand the latter measures lead to a considerable increase of the lateral field, which causes a number of limiting (4) or even detrimental hot carrier effects (5,6). As a solution, the operating voltage has to be reduced as well. This ideal scaling scheme is known as the constant field approach (7).

In reality, however, circuit and system level constraints have favoured a constant bias voltage over many generations of ICs The dominance of TTL devices has currently set the supply voltage standard at 5 Volt ± 10%. In addition, voltage margin requirements dictate that the above voltage cannot be scaled down at the same rate as the device dimensions. Consequently, the improvement of driving power has followed a law quite different from ideal scaling. This is shown in Fig. 1 for a number of process generations, characterized by a minimum gate width L . Rather than of being a constant (with bias voltage scaled) or a square law (with bias voltage constant), the driving current per unit gate length has improved at an intermediate rate. Near the 1µm level a slow-down occurs caused by velocity saturation effect (n-channel mainly). Deeper in the submicron range a further slow-down is observed due to surface roughness effects on mobility (8) and loss of effective gate driving capacitance (9). Furthermore, detrimental hot carrier effects make it unlikely that the bias voltage can be maintained at 5 Volt. Although these effects are interesting from a physical point of view, several other high-field effects are currently of primary concern because they affect device reliability.

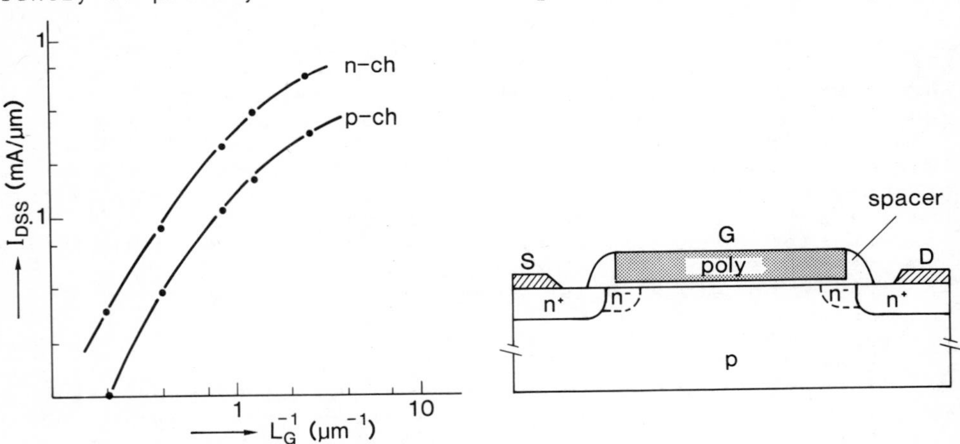

Fig. 1 Maximum current per unit gate width for 5 generations of Philips CMOS processes.

Fig. 2 Cross-section of LDD MOSFET.

In high-field operation two phenomena have to be distinguished: wear-out in thin-gate insulators and injection of hot carriers into the gate dielectrics. The first effect is associated with moving impurity ions or trapped charges. They cause a build-up of local fields in the insulator, leading to time dependent breakdown (10,11). Generally this effect can be kept under control by keeping the field in the insulator below 3 MV/cm. This limit is important for the development of small dynamic-RAM cells, but is of concern in active transistors only when approaching the 0.3 µm size. However, in the whole submicron range injection of hot carriers is the dominant effect.

When the applied lateral field in the channel becomes larger than 0.1 MV/cm, an appreciable number of carriers gain sufficient energy to surmount the gate insulator potential barrier and are injected by thermionic emission into the insulator (12,13). When the gate driving voltage is also high most of the injected carriers are able, due to the attracting field, to reach the gate. Therefore normally at $V_{gs} = V_{ds}$ a peak in the gate current is observed. However, the latter bias condition is rarely met in circuit applications. In fact a more serious situation occurs at low values of V_{gs}, when a much higher peak field is reached at the drain side of the channel. Then many more carriers are injected into a narrow area of 0.1 µm, but, since the field in the insulator has been inverted, most carriers never reach the gate and return to the silicon. Nevertheless, the abundant injection causes generation of interface states. Usually at a stress condition $V_{gs} \approx 1/2\ V_{ds}$, when the hole substrate current passes through a maximum, the above generation reaches a maximum also. Generally the states thus formed cause a long term change of threshold voltage or transconductance. Since the injection is an exponential function of the inverse of the field, the device lifetime (which is usually related to a fixed change of V_t or g_{ms}) appears to be an exponential function of the inverse of the applied drain voltage (14). This fact allows one to speed up lifetime tests or to evaluate hot-carrier-resistant devices or higher quality dielectrics.

In order to improve the lifetime or to extend the reliable voltage range, refined junction profile structures have been developed, which are known as LDD or DDD devices (15,16,17). Fig. 2 gives an LDD cross section. This approach reduces the peak field near the drain. This is

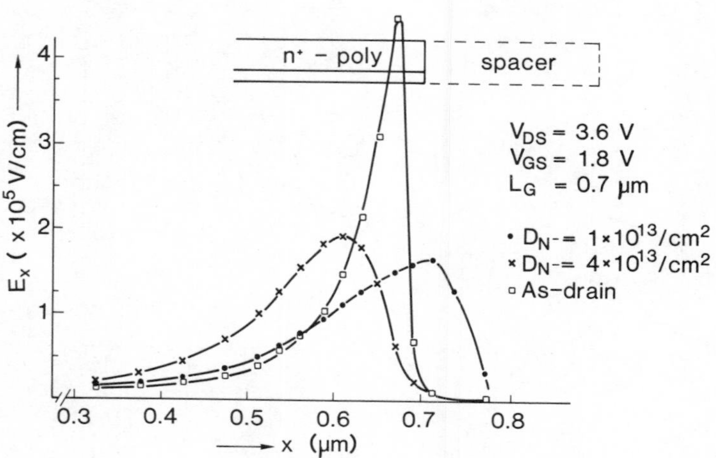

Fig.3 Field distribution along a current path for a conventional MOSFET and a Lightly Doped Drain (LDD) MOSFET, gate length 0.7 µm.

shown in Fig. 3 for an LDD device with a gate length of 0.7 µm. The lateral field distribution was calculated with a full 2-D device simulator (18). Because of the field reduction a considerable decrease of detrimental effects has been achieved. Generally both approaches can be optimized to produce similar results, but the DDD approach is more vulnerable to spacer width spread. Although a larger field reduction can be achieved at LDD doses lower than those shown in Fig. 3, usually the lifetime is worse. This fact can be understood from the results of Fig. 3. At decreasing LDD dose the location of the field peak gradually shifts to the heavily doped drain and towards the interface. Therefore the interface states are mainly filled, although at a reduced rate, in the lower quality spacer dielectrics. Calculations have shown that even a modest oxide charge is able to deplete the LDD section, causing a change in conductance (19). An additional disadvantage of applying a low LDD dose is the considerable loss of current driving capability owing to a relatively high series resistance at the source side. At high LDD doses the peak field gradually increases. However, due to the fact that the location of this peak moves towards the bulk, leading to weak avalanche generation further away from the interface, an optimum lifetime does not coincide with the lowest possible peak field or minimum substrate current (20). Furthermore, the above optimum depends on post diffusion procedure and junction depth. A typical stress result is shown in Fig. 4 (21). The resulting change in current drive applies to a 0.75 µm channel length. The extrapolated lifetime at V_{ds} = 4.5 Volt is estimated at 100 years.

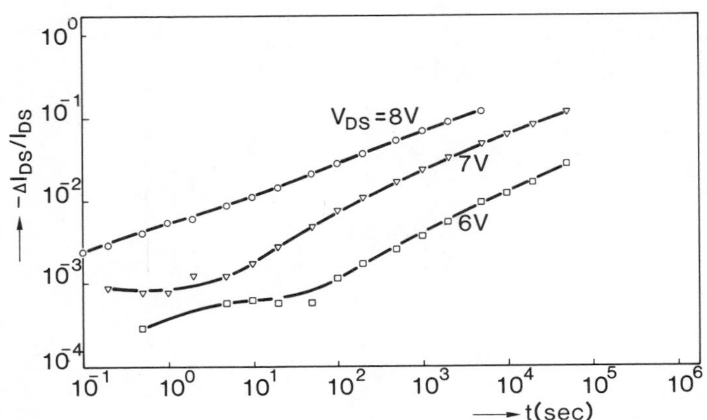

Fig. 4 Change of current versus time for a .75 µm n-MOST during operation at different drain bias.

In addition to the basic LDD approach of Fig. 2, a number of improved structures have been proposed recently. These includ applying an additional deep LDD implant (22), making use of a buried channel (23) and applying local anti-punch through implants (24). It is questionable, however, whether the claimed advantages with respect to lifetime balance the increased process complexity or other negative effects.

Generally at LDD doses yielding an optimum lifetime the series resistance effect on current drive is rather small. Nevertheless this effect, which is most likely caused by current crowding at the vicinity of the channel (25) and depletion of the LDD region, has to be taken into account in a circuit MOSFET model (26).

P-CHANNEL MOSFET

Although the p-channel MOSFET, as regards the physics of channel formation and operation, is basically no different from the n-channel type, owing to its construction, some properties differ. In CMOS technology p-channel MOSFETs are usually realized using a phosphorus-doped polysilicon gate. In order to obtain a practical threshold voltage between -0.7 and -1.0 Volts, in this case a shallow p-type layer is implanted in the channel region (compare Fig. 5). At the onset of conduction the depletion charges of the above layer and the adjoining substrate almost compensate each other. Current transport is by surface accumulation rather than by buried channel (27), and

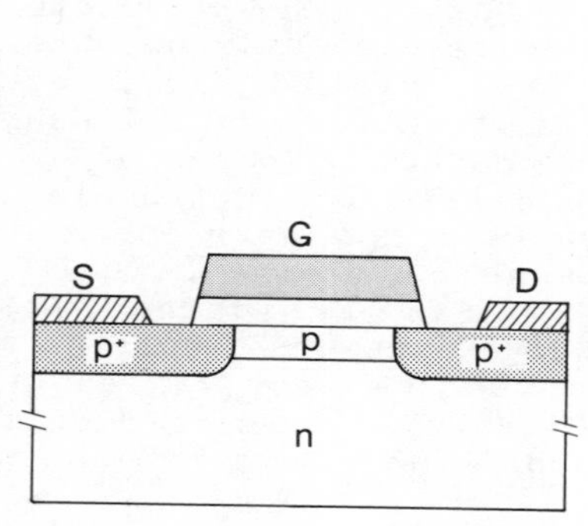

Fig. 5 Cross-section of p-channel MOSFET.

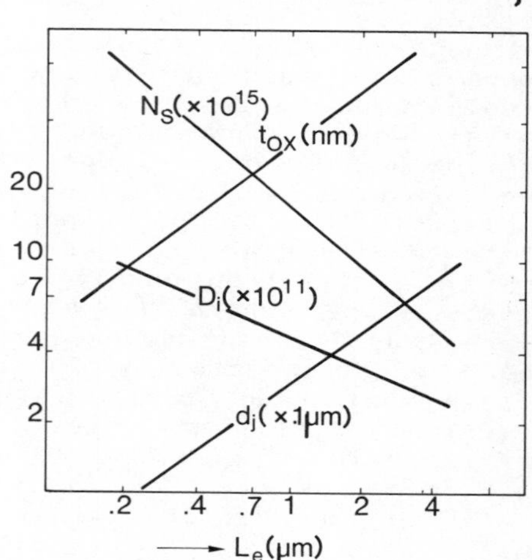

Fig. 6 Scaling rules for p-type MOSFET.

it is more appropriate to call this type of device a compensated MOSFET. As an additional advantage the above configuration has a threshold voltage, which hardly depends on the gate insulator thickness. The channel mobility which is relatively high (27,28). A disadvantage is the increased sensitivity to punch-through effects, which causes the subthreshold slope of submicron-channel devices to decrease considerably with drain bias. Essentially the latter effect is due to the fact that the potential barrier between channel and source has a minimum located in the bulk (approximately in the middle of the implanted p-layer) rather than at the interface. Therefore it is affected by field lines originating from the drain (29). In order to cope with the above problems the use of a p-type doped gate has been proposed (30,31,32). Although such a measure certainly cures the increased off-state current level, additional process complexity problems have to be faced. Therefore, from a practical point of view it is better to control the punch-through effect in the conventional p-type MOSFET by applying appropriate scaling rules.

Owing to the different threshold voltage dependence and the fact that hot carrier effects are much lower than for n-channel devices, different scaling rules would in fact be applicable. However, in order to be compatible with n-channel scaling, this degree of freedom is partially lost. Therefore, for practical scaling of p-channel devices only the punch-through effect is important. In devices with an effective channel length above 1 µm the potential barrier can be calculated from a 1-D solution of Poissons equation (27). Barrier lowering in submicron devices is essentially a 2-D effect which originates from the sideways pulling-down action from field lines terminating at the drain. Although it is intuitive that this effect has to be suppressed by measures aimed at forcing the potential barrier saddle-point to be located closer to the gate or at reducing the sideways action, a quantitative insight can only be obtained from a 2-D numerical calculation. As an example Fig.7 gives the potential barrier distribution as a function of well doping applicable to a 0.5 µm channel MOSFET operating in the off-state condition at V_{ds} = -6 V. The above distribution is taken in a direction normal to the surface and located at the saddle point, where the barrier reaches a minimum. While an average background doping level of 2.5 E16/cm^3 is too low to reach a satisfactorily low off-state current, applying an anti-punch-through implant proves to be an effective means of reducing the barrier lowering effect. With increasing dose the barrier height is increased and in addition the punch-through path is narrowed.

From a number of similar calculations it is concluded that although via the compensating p-layer dose a strong relation exists between threshold voltage and barrier height, the drain bias dependence is not affected. In fact decreasing the gate insulator thickness and drain junction depth, and increasing the well doping are more practical means of reducing the barrier-lowering effect (29). Generally, from the interpretation of the above simulations, a number of scaling rules can be derived aimed at securing a satisfactorily low value of the off-state current (33). These rules have been summarized in Fig.6 . Where a desired junction depth dj, a well or substrate doping level N_s, a threshold adjustment im

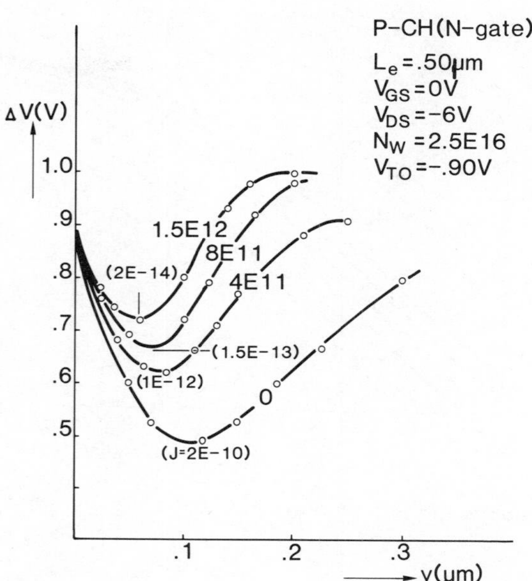

Fig. 7 Source-channel potential barrier distribution for different APT implants with off-state current level indicated.

plant D_i and a gate insulator thickness t_{ox} are given as a function of the effective channel length L_e. Of course in practice the above rules cannot always be maintained. For instance a source junction depth of 0.2 μm, which is required for an ideal scaling of a 0.5 μm channel device, is hard to achieve with present boron-based junction formation techniques. As a compromise it is possible to compensate the negative effects of a deeper junction by applying a higher anti-punch-through dose (compare Fig.7). In this way p-type MOSFETs with a channel length 0.5 μm and a junction depth 0.35 μm have been realized (34), which have an off-state current level of less than 1 pA/μm at V_{ds} = -4 V (compare Fig. 8). Naturally this is not an ideal situation. For instance the above result is obtained at the cost of an increased body effect which becomes manifest at longer channel lengths. Furthermore the strong dependence of the off-state current level on effective channel length makes it necessary to design the device for worst-case process conditions. In principle the increased body effect can be reduced by applying local punch-through stops in spacer technology (35). However, the required extra dose would be likely to increase the junction capacitance.

It is questionable whether the compensated structure can be successfully scaled down to a quarter micron gate. The scaling rules presented in Fig.6 require process innovations, but they may conflict with the fabrication of reliable n-channel devices of this size. In fact a much better compromise will be possible when a reliable gate material becomes available with a work function close to silicon mid gap (36).

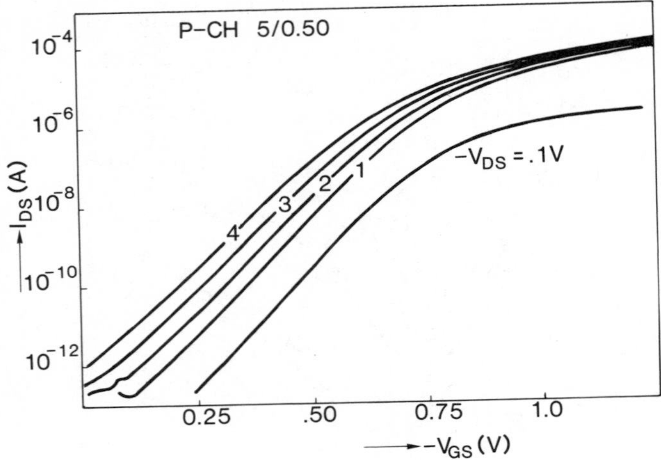

Fig. 8 Subthreshold characteristics of .5 μm p-channel MOSFET.

DYNAMIC RAM CELL

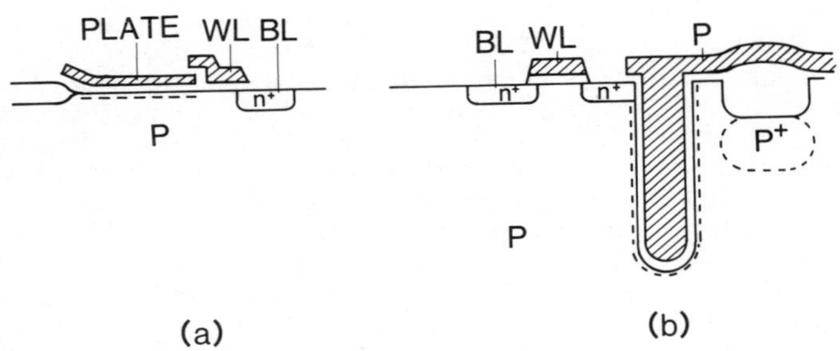

Fig. 9 Cross-section of planar (a) and trench (b) d RAM cell.

For more than a decade the dynamic RAM has been based on the 1-transistor cell concept shown in Fig. 9a. In this cell the information is carried by a charge Q_S, stored on a capacitor C_S (37). The lower electrode of C_S is either an inversion layer or an implanted region isolated in a substrate of the other type. The upper electrode is typically a polysilicon layer on top of a thin insulator. An additional MOSFET is used to switch C_S to a bit line. Upon read-out the signal charge (Q_S or 0) and a reference charge (1/2 Q_S) are dropped into a pair of bit lines. The voltage to be sensed is then (38) $V_S \approx 0.4\ Q_S/(C_b+C_S)$, where C_b is a parasitic capacitance associated with the bit line. Unfortunately, when expanding the memory matrix to mega-complexity, C_b, [which is already an order of magnitude larger than C_s] and V_s are subject to limitations. Owing to supply voltage variations and spread in circuit parameters V_s cannot be made lower than 100 mV. Because of non-scaling of spreading capacitance, C_b can hardly be smaller than 1pF/cm. Hence for mega-RAMs Q_s has to be kept at a value of 150 fC. An additional reason for the above choice is the impact of α-particles, leading to a possible charge collection of 30 fC (39).

In scaling down the cell in planar technology the charge required limits the d-RAM to the 1 mega-bit level. Considering a practical storage area of 12 μm^2 in this case, the simple MOS capacitor would contain fields of 4 MV/cm. By using an improved concept like the Hi-C cell (40) the latter value can be reduced to an acceptable value of 2 MV/cm. However, from the viewpoint of reliability for the 4M d-RAM, a different concept is needed. In order to enhance capacitance on a given chip area, stacked capacitors (41) based on high-ε dielectrics and folded cells (42) have been proposed. A successful realization of the latter type is the trench cell (43) shown in Fig. 9b . Since this cell is reasonably compatible with MOS technology and an increase in capacitance by an order of magnitude can be achieved, this concept is at present being developed intensively.

Since in trench-type cell operation a relatively high potential is present deep into the bulk, interference and punch-through between adjacent capacitors are limiting factors (44). This is illustrated in Fig. 10, where the measured leakage current vs distance between two 4-μm deep trenches has been plotted for various p-wells, realized by boron implantation and drive in(45). Unfortunately this result does not agree with simulation results. Although 2-D calculations with a simplified structure reveal a current flow path at half the depth of the trench (44), the simulated results are far too pessimistic. It has been suggested that owing to 3-D effects, penetration of substrate bias into the moat-to-moat gap causes a lifting of the potential barrier. No 3-D calculations have yet been published. Another punch-through path is possible between bit line junction and trench cell. From 2-D calculations it has been concluded that this becomes a problem when the related distance is reduced to 1.5 μm (46). Inter-trench punch-through is absent in a recently proposed configuration, where a heavily doped substrate forms the lower storage cell electrode and the pass-gate transistor is formed along the trench side-wall in an epitaxial layer (47).

Since the reporting of α-particle-induced soft errors in d-RAMs (48), efforts have been made to understand the physical mechanism and to develop an immune cell. However, the statistical character of the impact and the small value of induced charge complicate experimental studies considerably.
Numerical simulation has contributed more to a better understanding. Initial calculations were made by incorporating a realistic g-r model for α-injection into a 2-D device simulator (49), but recently the results obtained have been corrected by using a full 3-D simulator (39). From these studies some prac-

tical conclusions may be drawn:

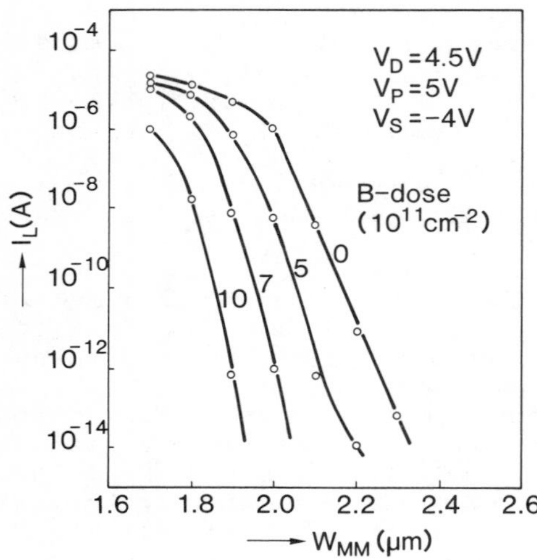

Fig. 10 Leakage current suppression by boron implanted p-well.

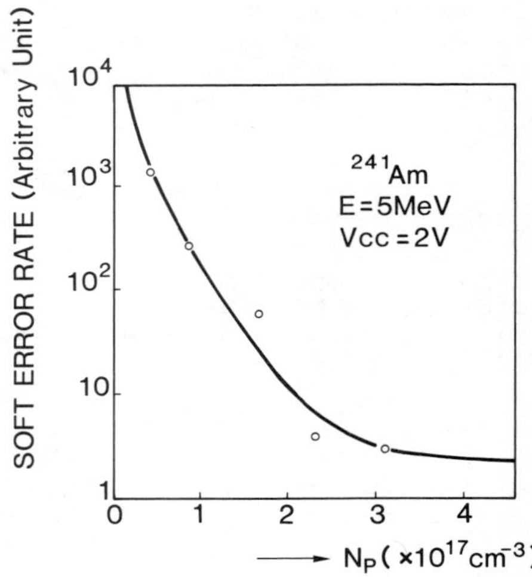

Fig. 11 Measured soft error rate versus well concentration.

- In contrast to physical intuition placement of the storage cell in a separate well is helpful. Owing to ambipolar effects on electron and hole diffusion of a few million generated pairs, a potential dip (50) along the particle trace occurs about a few picoseconds after impact.
- Since the collecting action of the storage mode is extended deep into the bulk, almost all generated electrons can move towards the Si surface (in less than 100 ps).
- Only by increasing the doping considerably can the above funnelling effect and consequently the collected charge be reduced. In practice the latter result has been achieved by implanting a buried layer in the well. Fig. 11 (51) shows the resulting soft-error rate as a function of peak concentration in the well. Although a much higher α-immunity seems feasible, it should be realized that owing to the high peak value the bit line capacitance is increased as well. Furthermore, the above results only apply to a planar cell. As yet no data for trench cells have been reported.

STATIC RAM CELL

In present static RAMs the memory cell is formed basically by six devices (a cross-coupled flip-flop and two pass-gate transistors). Historically, most of the improvements in cell packing density have come from innovations in the flip-flop load devices. Most present designs use high-ohmic polysilicon layers placed on top of the MOSFET switches. Defects in the grain boundaries give rise to trapping states that reduce the number of free carriers and create space-charge regions in the crystallites and potential barriers that impede carrier motion. Carrier transport and therefore the resistivity are mainly determined by thermionic field emission of carriers across a barrier (52). Two typical plots of resistance behaviour are given in figs. 12 and 13 (53). At low doping rates the resistance

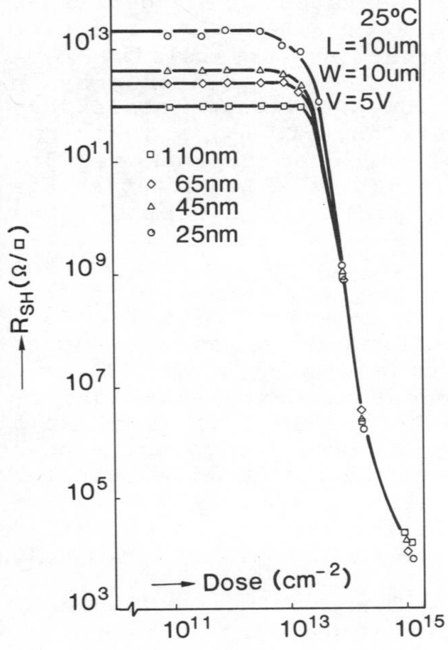

Fig. 12 Sheet resistance versus arsenic dose for polysilicon resistors.

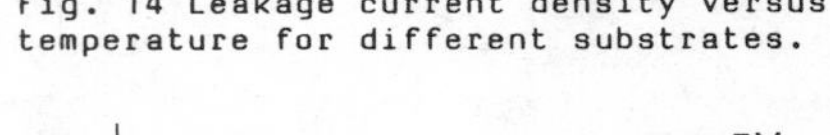

Fig. 14 Leakage current density versus temperature for different substrates.

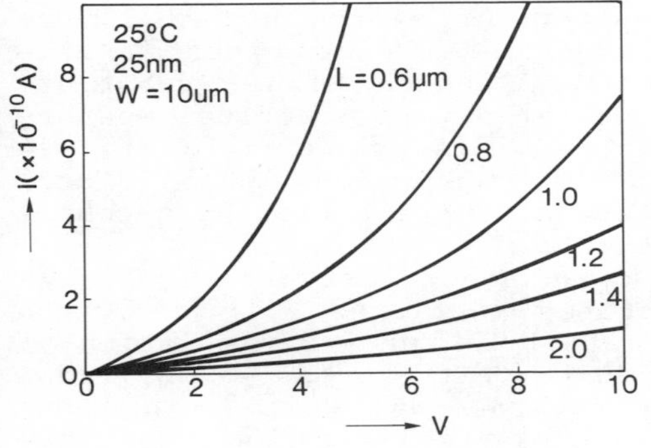

Fig. 13 I-V characteristics of short polysilicon resistors.

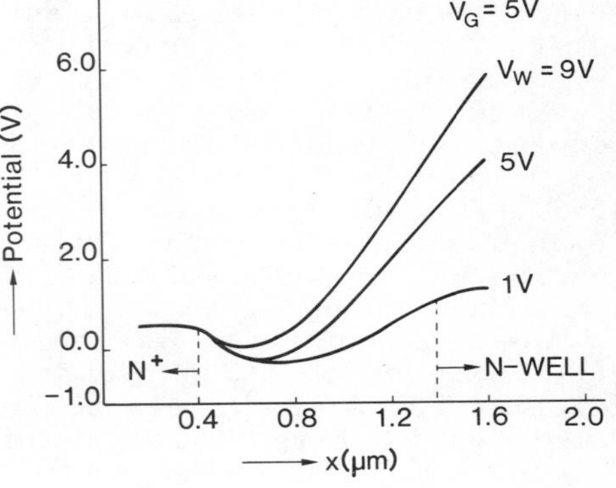

Fig. 15 Potential distribution in n+-source to n-well field transistor.

is completely determined by the grain boundaries. Since the average crystallite dimension (L_g) is related to film thickness (53), in thinner films there is less voltage across each grain and therefore thermionic emission is reduced. When the doping rate N approaches a value $NL_g = N_t$, where N_t is the number of traps in the grains (usually $10^{13}/cm^2$), the traps become filled and the barrier height is reduced. This corresponds to a sharp decline in resistivity. Fig. 13 illustrates the thermionic emission effect on the I-V characteristics. When the resistor length is shortened, the applied voltage per crystallite $V_g = V/N_g$ is increased and for $V_g \sim 0.10$ V the resistance has a large voltage coefficient. Since the low-doped layer resistance is too high, present RAMs use load devices with $R_{sh} = 10^9 \Omega/\square$, which according to Fig. 12 have a strong dope dependence. The associated large spread of these loads mainly affects the stand-by current and soft-error rate, but not the access times of s-RAMs.

When approaching mega-complexity a tighter control of the load devices is required. Since the overall resistor dimensions have to be reduced and current transport via a few crystallites has to be avoided, film thickness has to be reduced and relatively large lateral diffusion over several µm from the contact diffusion has to be eliminated. Next, although according to Fig. 13, low-doped loads with currents of 10 pA at 5V can be achieved with W = L = 1 µm (leading to a quite acceptable stand-by current of 10 µA) one should realize that for low V the resistance may increase by almost an order of magnitude. Because of this and particularly in view of the spread in R_{sh}, the leakage current of the MOSFETs must be lower than 0.10 pA.

As an alternative to poly load s-RAMs, a full CMOS RAM or a pseudo-static RAM may be considered. In order to save area very small transistors and narrow isolation lines (see next section) are needed for the first type. For the other type, which is basically a d-RAM with refresh circuitry integrated on chip, dissipation can be reduced by decreasing the refresh rate. The latter is made possible by reducing the minority carrier generation rate. In present, uniformly doped substrates the diffusion length is several hundreds of micron and therefore with small isolated junctions edge effects can increase the leakage current considerably (54). Although higher substrate doping may reduce the generation rate, a more practical measure is to use $p\text{-}p^+$ epitaxial substrates. This is shown in Fig. 14 (55). An additional advantage is the reduction of edge effects owing to strong suppression of lateral current flow by recombination in the heavily doped substrate.

Unfortunately not much is known about soft-error rates in s-RAMs. In addition to the problems mentioned with d-RAMs a realistic calculation requires a full 2-D or 3-D simulation of carrier transport in the semiconductor with boundary conditions dictated by flip-flop circuitry environment. Considering that according to the d-RAM results an α-particle may cause collection of 10-30 fC of charge on an isolated junction within 100 ps (39), it is realistic to assume that a poly load cannot transfer this charge in that time. Therefore, cell junction capacitance has been enhanced considerably . It is questionable whether this measure is yet practical. Considering that a 2.0/0.7 μm p-type MOSFET biased at V_{gs} = 4 V and V_{ds} = 1 V may carry a current of 0.25 mA with a transit time of 25 ps, it seems feasible that for mega-complexity a full CMOS cell could handle the charge pulse from the impact of α-particles.

ISOLATION

Owing to the inherently smooth transition between active area and isolation regions, and the self-aligned field implant, local oxidation (56) has been the standard isolation technology for about a decade. Remarkably, the dominance of the original LOCOS has become an obstacle in full process down-scaling. Generally three problems arise. First a wasted space of up to 1 μm (57) is lost to the gate insulation region (called birds beak). Next comes a loss of useful space because of lateral diffusion of the field implant (called field encroachment). Usually the latter effect causes an increase of threshold voltage and other narrow channel width effects. Finally, in CMOS a self-aligned field implant can only be partially achieved and at the cost of a high-doped n-type region (compare the insert of Fig. 16 for an n-well process). As a result the n^+ - p^+ junction spacing cannot be reduced at will. Owing to this fact and to the poor scaling of interconnection capacitances the vertical and lateral dimensions of LOCOS patterns have been reduced at a much lower rate than other dimensions.

As a radical solution to overcome the above problems, trench-type isolation schemes have been proposed (58, 59). Basically these techniques consist of cutting a hole in the substrate and refilling it with an insulator. However, the surface planarity is basically lost, the formation of the insulation region is rather complicated and the potentially larger sidewall inversion requires higher doped n- and p-type surface layers. This solution has not yet matured (60).
Therefore, from the manufacturing viewpoint, it is more attractive to improve LOCOS. Proceeding in order of complexity, birds beak has been reduced by etching back (retaining a fairly large bias from mask to final device edge), by using oxide-nitride films to reduce lateral oxidation (61) (leading to a less than flat surface) and by framing the active area edge to suppress lateral oxidation almost completely (62). Recently after better insight had been gained into the nonuniform stress caused by volume expansion accompanying the oxidation process, the same results were obtained by applying a more sophisticated oxidation procedure (63). These facts and the higher punch-through requirements in scaling-down processes leading to fewer encroachment problems, demonstrate that the previously mentioned shortcomings of original LOCOS can largely be remedied.
In fact only the limitations in n^+ - p^+ spacing remain. This is shown in Figs. 15 and 16. Fig. 15 shows, for instance, that the potential barrier between shallow n^+ junction and channel stop area is lowered by an increased well potential. In this 2-D si-

mulation result (64) care was taken to use realistic rounded LOCOS profiles to avoid too optimistic results associated with corner effects (65). Fig. 16 gives the calculated worst-case leakage currents as a function of critical spacing. Taking into account mask tolerance these results show that a minimum n^+ - p^+ spacing of 3.0 μm is feasible with modified LOCOS. A considerable reduction of this spacing is possible if the conventional field implant is replaced by implanting after oxide formation (66).

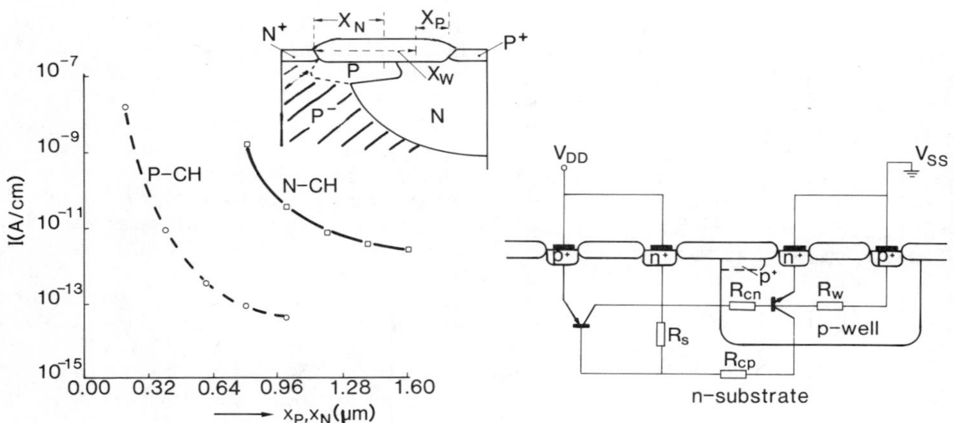

Fig. 16 Field transistor leakage current versus relevant mask distance with cross-section of modified LOCOS as inset.

Fig. 17 Cross-section of p-well CMOS with corresponding thyristor circuit.

Any bulk CMOS technology inevitably incorporates a four-layer n^+ - p - n - p^+ path between the supply rails. Fig. 17 gives an overview for a p-well case. Although the current gain product of both bipolar transistors is larger than one, under normal bias conditions the critical voltage is so high that the parasitic thyristor remains in the high impedance state. In practical circuit configurations triggering of latch-up is most likely to occur through forward biasing of the BE junction of the high-gain well transistor. This effect may arise through a number of mechanisms. For instance, electron-hole pair generation by weak avalanche multiplication in a nearby n-channel transistor or dynamic overshoot of a p^+ junction in the substrate, causes holes to flow along the well to the nearest supply contact and to forward bias an n^+ source (compare fig. 17). Alternatively, during power-up the well substrate capacitance may not charge fast enough to keep the well at the same potential as an n^+ source, and latch-up is initiated as soon as the device is turned on (67).

Since the current path before and after firing of the thyristor has completely changed and high-injection and 3-D effects play a role, an exact calculation of the thyristor characteristics or a forecast of latch-up conditions is only possible via numerical analysis (68). As an example in

Fig. 18 a sequence of potential distributions during power-up are given (70). The injecting current causes a potential bulge

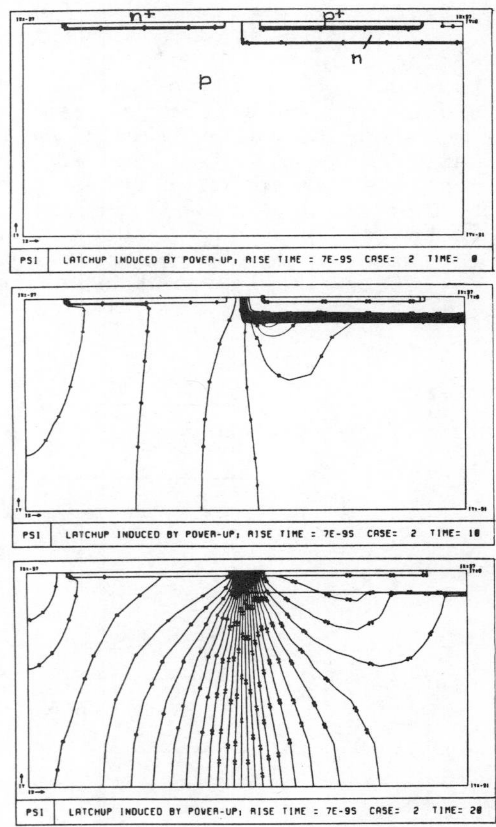

Fig. 18 Potential distribution in n-well CMOS structure at t=0, 10 ns and 20 ns during power-up (dimension along x-axis 100 μm).

away from the well followed by a firing of the thyristor. Owing to limited computer time and a limited number of grid points, only elementary configurations, neglecting circuit surroundings, have been studied in this way (68, 69). Consequently few design rules have been obtained.

In order to circumvent the above problem, a number of lumped element models have been proposed (71, 72, 73). A minimum model is found in Fig. 17 and is made up of two transistors, two base resistances R_W and R_S, and two collector resistances R_{cn} and R_{cp}. The well resistances are not difficult to calculate, Owing to 2-D or even 3-D spreading effects the others can only be obtained from numerical analysis (74) or conformal mapping techniques. Following this approach a first order calculation is possible of the critical current to initiate latch-up (75). However, the accuracy in calculation holding current or voltage is questionable, since in this case transistor high injection effects (69), sharing of collector current over several substrate contacts and emitter resistance (76) also play a role. In this respect some improvement is possible by using a mixed device-circuit simulator (77, 78). However, errors in excess of 100% are possible.

In spite of this situation a number of useful conclusions can be drawn:

1. Guard rings placed as close as possible around injecting diodes are an effective means of reducing latch-up, in particular in combination with the use of epitaxial layers (79). However, this solution is generally only possible for input or output sections.

2. Increasing the well doping and replacing bulk material by epitaxial substrates causes an increase of I_{cr}, I_h and V_h. This is shown by the survey in Fig. 19, where original data (80) have been supplemented by recent data. The difference between

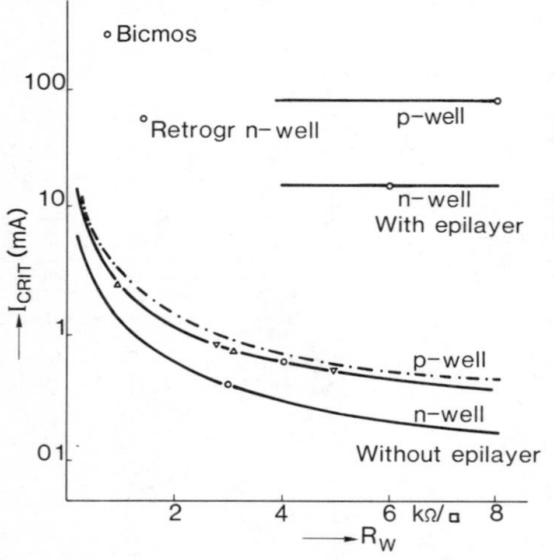

Fig. 19 Critical current versus well resistance for different CMOS technologies.

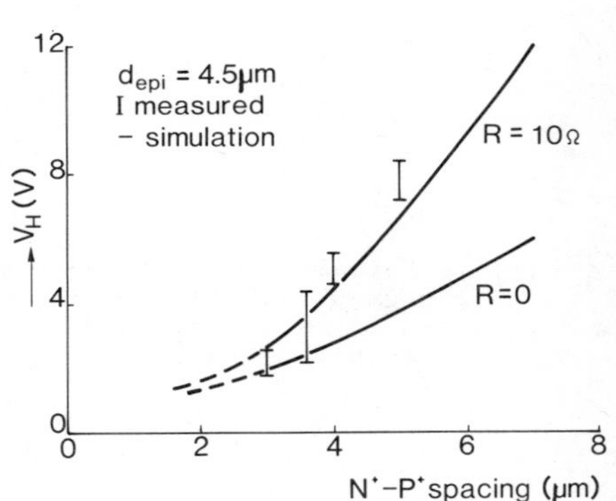

Fig. 20 Holding voltage versus n^+ source to p^+ source spacing.

n-well and p-well epitaxial processes mainly arises from the thicker epi-layers needed to accommodate the larger updiffusion of a p^+ substrate. By applying shallow n-well junctions (81) or retrograde wells (76) the same results as for p-wells were later obtained. A BIMOS process with two updiffused wells is even better (82).

3. Owing to the decrease of collector resistance, a reduction of $n^+ - p^+$ spacing causes a considerable decrease of I_h and V_h. Nevertheless by scaling the epi-layer with the above spacing a holding voltage larger than 6 V has been realized at a spacing of 4 µm (76) (compare fig. 20).

4. Owing to a reduction of base resistance the use of reverse and in particular butted substrate and well contacts can be an additional means to improve latch-up hardness (83).

5. Satisfactory improvement of latch-up immunity at $n^+ - p^+$ spacings of 1 µm has been reported by using trench isolation

(84, 85). However, effective killing of the parasitic thyristor is only possible by using complete substrate isolation.

INTERCONNECTION

Since the dimensions of interconnecting wiring more or less have to follow the shrinking dimensions of active devices, several problems arise. Owing to the increased current density electromigration in the wires and contact areas may cause malfunctioning of the IC. In addition the increased resistance of small contact areas may degrade circuit performance. In principle the above deficiences can be solved by using additional metal (86, 87) to aluminium and by applying salicide layers (88, 89) to reduce the potential barrier at the contact In fact, the relevant problems have more to do with process engineering than with physics. However, owing to topological constraints and to the fact that the thickness can less easily be scaled down , the wiring needed to connect submicron devices can no longer be viewed as ideal parallel plate capacitors.

As a result, a large capacitive coupling is formed between wires and from wires to devices, and the circuit performance deteriorates. The problem is aggravated by the fact that adequate test structures cannot always be designed. Therefore it is faster to calculate capacitances.

Since the coupling is 2- or 3-dimensional such a calculation is complicated. Although several approximate formulae for various configurations have been given (90, 91, 92), their value is limited. In practice a topology is often used, which is too general to adapt to the simplified schemes of analytic solutions. Therefore numerical routines based on solving Poisson's equation are needed to calculate the capacitances between 3-D arbitrary configurations (93). This will be demonstrated for two cases (94).

The first configuration shown in the inset of Fig. 21 is formed by three parallel conductors running on top of a substrate. In order to understand the capacitance behaviour we first consider the simple case of one conductor (F) and a substrate. When the width (W) of the line is not large compared with the distance (H) and the thickness (t) of the conductor has a comparable size, the line to substrate capacitance strongly deviates for several reasons from the 1-D case. Owing to charge crowding the field peaks under the lower corners of the conductor in a direction that is not normal to the substrate. Furthermore charge from the sidewalls and the top surface contribute to the capacitance. As a result, for W = t = 1 μm the 2-D capacitance is a factor of 3 to 5 larger than the 1-D case for H = 0.6 μm and H = 1.0 μm, respectively. An approximation, which gives the specific contribution from various parts by Schwarz-Christoffel transformation (90) has been found to deviate from numerical results by only 10%. This is no longer true for the general case of Fig. 21. When the coupling distance S is decreased, the capacitance C_{FB} is reduced. This is caused by shielding of field lines coming from

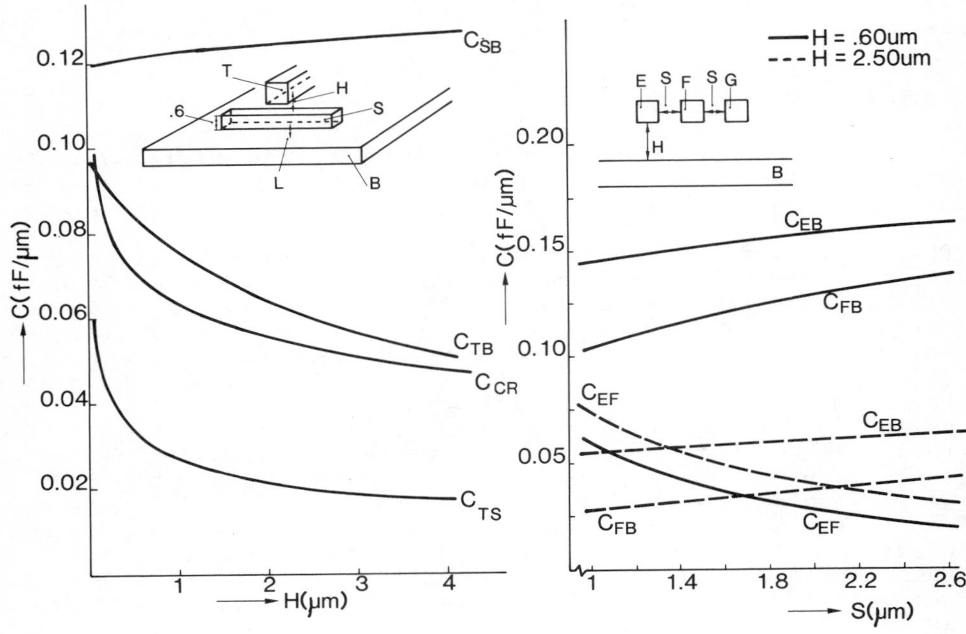

Fig. 22 Capacitance between two crossing lines and the substrate.

Fig. 21 Interline capacitance versus line spacing.

the sidewalls and the top surface, and a reduction of the normal field below the conductor F by the adjacent conductors. To a lesser extent the same arguments apply to the behaviour of C_{EB}. On the other hand with decreasing S the coupling capacitance C_{EF} increases. Here the substrate has a shielding effect on the value of C_{EF}. An approximation has been given only for the total capacitance of F (92), and has been found to deviate by as much as 25% from the numerical results. Other configurations like two conductors at a different height above the substrate and partially overlapping show similar subtle effects (94).

The second configuration is formed by two conducting lines forming a cross above a substrate (shown in the inset of fig. 22). Owing to charge crowding effects and strongly nonuniform field distributions, and to shielding effects on the contribution of sidewalls and top surface, the various capacitances cannot be obtained by considering the 3-D problem as a sum of separate 2-D cross-sections (95). In Fig. 22 the result of a 3-D numerical calculation for the different capacitances is given as a function of the distance between the two lines. The cross-section of the conductors in 1 x 1 μm^2 (for T) and 1 x 0.6 μm^2 (for S). The distance between S and the substrate (B) is 1 μm. Increasing the separation between the conductors decreases the coupling capacitance C_{TS}, but increases the error in estimating that capacitance with simple models.

For instance, at H = 1 μm, C_{TS} is 9 times larger than the 1-D overlap capacitance. Furthermore, increasing H by 100% only decreases the coupling capacitance by 30 %. When the substrate is omitted, C_{TS} increases further to C_{CR}, as the substrate no longer causes a shielding of field lines from S. The increase of C_{SB} with increasing H is also an effect of shielding (in this case due to the presence of conductor T).

In fact the two cases given above are still elementary examples of VLSI wiring. Present limitations in mesh generation do not allow us to calculate all effects in possible wiring configurations. Nevertheless the results are useful and underline the need for careful consideration of capacitive effects when designing VLSI circuits and technology.

CONCLUSIONS

Specific problems arising from scaling down of device dimensions to the submicron range have been discussed. In order to understand and cope with the above problems a combination of relevant experiments and 2-D or 3-D device simulations has been indispensable. Although most effects will allow the realization of VLSI with megabit complexity, in the near future changes in materials and processing techniques are required.

ACKNOWLEDGEMENT

The author acknowledges the advice and help of his colleagues P.T.J. Biermans, R.J.G. Goossens, C.D. Hartgring, R. Petterson, P.A. v.d. Plas, T. Poorter, J.W. Slotboom, P.J. v.d. Wiel, P.H. Woerlee and the students P.H.A. Spijkers, J.H.M. Quint of the Eindhoven University.

REFERENCES

1 L.D. Yau, Sol.St. Electr. 17, 1059 (1974)

2 K.N. Ratnakumar et al, IEEE J. Sol.St. Cir. SC -17, 937 (1982)

3 R.R. Troutman et al, IEEE Trans.El.Dev., ED-24, 1266 (1977)

4 G. Merckel et al, IEEE Trans.El.Dev., ED-19, 681 (1972)

5 E. Takeda et al, IEE Proc. 130, 144 (1983)

6 F.C. Hsu et al, IEEE Trans.El.Dev., ED-32, 394 (1985

7 R.H. Dennard et al, IEEE J.Sol.St. Circ., SC-9, 256 (1974)

8 T. Ando et al, Rev. Mod. Phys. 54, 437 (1982)

9 Y.A. El-Mansy, Proc. ICCC 80, 457 (1980)

10 E. Harari, J. of Appl. Phys. 49, 2478 (1978)

11 D.R. Wolters et al, Series Electrophysics 7, 111 (1981)

12 T.H. Ning, IEDM Techn.Dig. p. 144 (1977)

14 E. Takeda, Symp. VLSI Techn., p. 2 (1985)

15 S. Ogura et al, IEEE Trans.El.Dev., ED-27, 1359 (1980)

16 S. Ogura et al, Techn.Dig. IEDM, p. 718 (1982)

17 E. Takeda et al, IEEE Trans.El.Dev., ED-30, 652 (1983)

18 CURRY - Philips proprietary 2-D device simulator
19 P.T.J. Biermans - ESSDERC conf. abstr., p. 312 (1985)
20 P.T.J. Biermans et al, ESSDERC 1986, paper C2
21 T. Poorter et al, to be published
22 H. Grinolds et al, IEDM Techn.Dig., p. 246 (1985)
23 N. Nakahara et al, IEDM Techn.Dig., p. 238 (1985)
24 C. Codella et al, IEDM Techn.Dig., p. 230 (1985)
25 G. Bacarrani et al, IEEE El.Dev.Lett., EDL-4, 27 (1983)
26 F.M. Klaassen et al, to be published
27 F.M. Klaassen et al, Sol.St.Electr. 28, 359 (1985)
28 E. Sun et al, IEDM Techn.Dig., p. 791 (1980)
29 F.M. Klaassen et al, IEDM Techn.Dig., p. 613 (1984)
30 L.C. Parillo et al, IEDM Techn.Dig., p. 418 (1984)
31 S.J. Hillenius et al, ESSDERC Conf.Abstr., p. 145 (1985)
32 M. Yoshimi et al, Dig. VLSI Symp., p. 76 (1984)
33 F.M. Klaassen, to be published
34 P. Woerlee et al, to be publis hed
35 S. Odanaka et al, IEEE Trans.El.Dev. ED-33, 317 (1986)
36 U. Shwabe et al, IEEE Trans.El.Dev., ED-31, 988 (1984)
37 R.H. Dennard, U.S. patent 3 387 286
38 K.U. Stein et al, IEEE J.Sol.St.Circ., SC-7, 336 (1972)
39 H. Masuda et al, Techn.Dig. IEDM, p. 496 (1985)
40 A. Tash et al, IEEE Trans.El.Dev., ED-25, 33 (1978)
41 M. Koyanugi et al, Techn. Dig. IEDM, p. 348 (1978)
42 J.C. Sturm et al, IEEE El.Dev.Letts, EDL-5, 151 (1984)
43 T. Mano et al, Techn.Dig. ISSCC, p. 234 (1983)
44 H. Sunami et al, IEEE Trans.El.Dev. ED-31, 746 (1984)
45 H. Sunami et al, Techn.Dig. IEDM, p. 232 (1984)
46 R. Kircher, to be published
47 H. Schichijo et al, IEEE El.Dev.Letts, EDL-7, 119 (1986)
48 T.C. May et al, IEEE Trans.El.Dev., ED-26, 2 (1979)
49 C.M. Hsieh et al, IEEE Trans.El.Dev., ED-30, 686 (1983)
50 C.M. Hsieh et al, IEEE El.Dev. Letts, EDL-2, 103 (1981)
51 H. Momosle et al, Techn.Dig.IEDM, p. 706 (1984)
52 J.Y.W. Seto, J. Appl. Phys. 46, 5247 (1975)
53 T. Ohzone et al, IEEE Trans.El.Dev., ED-32, 1749 (1985)
54 J.W. Slotboom et al, Techn.Dig. IEDM, p. 667 (1984)
55 J.W. Slotboom et al, IEEE El.Dev. Letts, p. 403 (1983)
56 J.J.A. Appels et al, Phil.Res.Rpts. p. 118 (1970)
57 W.G. Oldham, Techn.Dig.IEDM, p. 216 (1982)
58 R.D. Rung et al, Techn.Dig.IEDM, p. 237 (1982)
59 T. Shibata et al, Techn.Dig.IEDM, p. 27 (1983)
60 R.D. Rung, Techn.Dig.IEDM, p. 574 (1984)
61 J. Hui et al, IEEE El.Dev. LeTTS, EDL-2, 244 (1981)
62 K.Y. Chiu et al, IEEE Trans. El. Dev., ED-30, 1506 (1983
63 P. v.d. Plas et al, to be published
64 P.H.A. Spijkers, MSc Thesis-Eindhoven University (1986)
65 S.H. Goodwin et al, IEEE Trans. El.Dev., ED-31, 7 (1984)
66 R.A. Martin et al, IEDM Techn. Dig., p. 403 (1985)
67 R.R. Troutman et al, IEEE Trans. El. Dev., ED-30, 170 (1
68 A. Wieder et al, IEEE Trans. El. Dev., ED-30, 240 (1983)
69 M.R. Pinto et al, IEEE El. Dev. Letts, EDL-6, p. (1985)
70 R.J.G. Goossens, to be published
71 D. Estreich, Ph. D. Thesis-Stanford University (1980)
72 A.G. Lewis, IEEE Trans. El. Dev., ED-31, 1472 (1984)
73 R. Fang et al, IEEE Trans. El. Dev., ED-31, 1 (1984)

74 M.J. Chen et al , Sol. St. Electr. 28, 855 (1985)
75 M.J. Chen et al, Int. Symp. VLSI Technology, p. 240 (1985)
76 Y. Taur et al, IEDM Techn. Dig., p 398 (1984)
77 W.L. Engl et al, IEEE Trans. CAD-1, 85 (1982)
78 J. Harter et al, IEEE trans. El. Dev., ED-32, 1665 (1985)
79 R.R. Troutman et al, IEEE El.Dev. Letts, EDL-4, 438 (1983)
80 D. Takacs et al, ESSDERC Conf. Abstr., p. 108 (1983)
81 L.C. Parillo, IEDM Techn. Dig., p. 398 (1985)
82 BICMOS, Philips proprietary CMOS process
83 F.S. Lai et al, IEDM Techn. Dig., p. 513 (1985)
84 T. Yamaguchi et al, IEEE Trans. El. Dev., ED-32, 184 (1985)
85 Y. Niitsu et al, IEDM Techn. Dig., p. 509 (1985)
86 I. Arne et al, IBM J. Res./Dev. 14, 461 (1970)
87 J.K. Howard et al, I of Appl. Physics 49, 4083 (1978)
88 C.Y. Ting, IEDM Techn. Dig., p. 110 (1984)
89 T. Moriya et al, IEDM Techn. Dig., p. 550 (1983)
90 E.W. Greeneich, IEEE Trans. El. Dev., ED-30, 1838 (1983)
91 R.L. Dang et al, IEEE El. Dev. Letts, EDL-2, 196 (1981)
92 T. Sakurai et al, IEEE trans. El. Dev., ED-30, 183 (1983)
93 D.J. Coe et al, Nasecode III, p. 102 (1983)
94 J.H.M.M. Quint, MSc Thesis - Eindhoven University (1986)
95 P.E. Cottrell et al, IBM J. Res/Dev. 29, 277 (1985)

Inst. Phys. Conf. Ser. No. 82
Paper presented at ESSDERC 1986, Cambridge 8–11 Sept. 1986

Metalorganic molecular beam epitaxy (MOMBE)

H Lüth

2. Physikalisches Institut der Rheinisch-Westfälischen Technischen Hochschule Aachen, D-5100 Aachen, Fed. Rep. of Germany

Abstract. Metalorganic MBE (MOMBE) is a new technique for the controlled growth of thin III-V semiconductor films. As in MBE the process is carried out in an UHV system; as sources molecular beams of hydrides like AsH_3, PH_3 and/or metal alkyls (TMG etc.) are used. Different experimental approaches to MOMBE are discussed. The main emphasis is laid on growth and doping of GaAs. First results concerning the production of device structures are reported and the possibility of selective growth by MOMBE is considered.

1. Introduction

Progress in the field of very fast and of optoelectronic III-V semiconductor devices and integrated circuits is intimately related with an improvement of the techniques for epitaxy. Complex structures as they are used for quantum well LASERS, in fast bipolar logic circuits or in high electron mobility transistors (HEMT) require a high accuracy in the control of layer thickness, composition (for ternary and quaternary compounds), and of doping. Extremely sharp doping and composition profiles with high crystallographic quality on an atomic scale are necessary. For the production of integrated circuits these requirements have to be fulfilled in addition to that of a high degree of lateral homogeneity over a large area. A further condition for the production of integrated circuits is the availability of layers with extremely low surface defect density. The generation of surface defects during the growth is a severe problem e.g. in standard molecular beam epitaxy (MBE).

Three experimental approaches are widely used at present in the field of III-V compound epitaxy. In the metalorganic chemical vapour deposition (MOCVD) technique a cold wall flow reactor is used, into which the source gases, for GaAs epitaxy AsH_3, metal alkyls
$Ga(CH_3)_3$] and H_2 as carrier gas are injected. The process is performed at atmospheric pressure or under low pressure conditions (LPMOCVD).

In the molecular beam epitaxy (MBE) the growth process runs in an ultrahigh vacuum (UHV) system, where molecular beams as sources are generated by effusion from ovens (effusion cells) containing the corresponding solid elements (Ga, In, Sb, As etc.).

The obvious advantages of MOCVD, continuous performance even for large numbers of wafers and easy control of deposition rates and doping by electronic mass flow controllers, are contrasted in MBE by the possibility of easily producing extremely sharp doping and composition profiles by using mechanical shutters. Since no gas volume has to be changed, switching times between different source materials and dopants are below the time required to grow a monolayer.

The third technique, MOMBE, was developed in order to combine the advantages of MOCVD and MBE (Veuhoff et al 1981). In MOMBE an UHV system is used as growth chamber and, as with MOCVD, gas sources (AsH_3, PH_3, trimethylgallium TMG, triethylgallium TEG etc.) supply the growth material and also the dopant (SiH_4 etc.). Their flow can easily be controlled by valves from outside the UHV system. As in MBE, short switching times are achieved and a continuous performance is possible. Unlike MBE the source containers do not have to be refilled and irreproducibilities due to refilling and breaking of the UHV are avoided. The UHV approach in MOMBE furthermore allows growth control on an atomic scale by RHEED (reflection high energy electron diffraction) oscillations. The technique is also compatible with technological processes carried out in high vacuum, such as ion implantation ion etching etc.. This might save time in a production process.

The present review is intended to give a short overview over the different experimental approaches to MOMBE and to show the state of the art in this rapidly growing new field of III-V epitaxy. Based on the results obtained so far with MOMBE, there are special expectations that device technology in the fields of optoelectronics and fast devices and circuits (MESFET, HEMT etc.) can be improved considerably.

2. Several Approaches to MOMBE

The consideration that effusion cells for the group V elements in MBE are rapidly depleted and cause irreproducibilities and inhomogeneities of the layers led Panish (1980) to the use of gas sources of AsH_3 and PH_3 for the growth of GaAs, InP and the quaternary heterostructure GaInAsP/InP (Panish and Tempkin 1984, Tempkin et al 1985). The same approach, exchange of the group V element effusion cell with a gaseous AsH_3 source was applied by Calawa (1981) and Huet and Lambert (1986). Similarly for the growth of phosphorus compounds in a MBE system, a PH_3 molecular beam with cracking facility was used by Chow and Chai (1983).

The complementary approach, exchange of the element III effusion cell in the MBE apparatus with a metalorganic molecular beam source, was investigated by some Japanese groups (Tokumitsu et al 1984 and 1985, Kawaguchi et al 1984, Kondo et al 1986). Metalorganic compounds, e.g. TMG, TEG and TEIn, supply Ga and In to the growing surface, and also the effect of additional introduction of atomic hydrogen for reducing the incorporation of carbon (background p-type doping) was studied (Kawaguchi et al 1984, Calawa 1979).

The first consequent application of the MOMBE process, i.e. using gas line beams for both the group III (TMG, TEG) and the group V elements (AsH_3) was performed by our own group for the growth of GaAs (Veuhoff et al 1981). Somewhat later Vodjdani et al (1982) obtained similar results by means of a comparable approach. The applicability of two gas line sources (TEIn, PH_3) for the growth of InP was shown by Kawaguchi et al (1985). Recently, these authors succeeded in growing high-quality InP layers with a 77 K mobility of 105000 cm^2/Vs and a free electron concentration of $9.1x10^{13} cm^{-3}$ (Kawaguchi et al 1986). In order to achieve a good control of the film thickness, Nishizawa et al (1985) recently used pulsed gas line sources (TMG, AsH_3) to grow GaAs monolayer by monolayer in an UHV system.

An interesting version of the MOMBE approach with two gas line sources, called "Chemical beam epitaxy", is presented by Tsang (1984). He supplies both the group III and the group V component as metalorganics (TMIn, TEIn, TMG, TEG, TMAs, TEP). The quality of the layers as far as surface defect density and p-type background doping are concerned, is improved by additional mixing of H_2 into the gas line (Tsang 1985a) and by using TEG (Tsang 1985b), respectively.

If one compares the different experimental apparatus for MOMBE which are in use so far, the major difference consists in the cracking facility for the element V gas line. While the inlet capillary for the metalorganic compound has to be heated only slightly above room temperature in order to avoid condensation due to the adjacent cryoshield, the gaseous group V starting materials with high thermal stability have to be precracked in the injection capillary. Most groups including our own use a so-called low pressure cracking capillary, where the AsH_3, PH_3 etc. is thermally decomposed by means of a heated metal filament (mostly Ta at $\sim$1000 K) (Fig. 1a). Typically, the AsH_3 gas is injected into the UHV system by a controllable leak valve, which allows a reproducible adjustment of the beam pressure, i.e. also of the flux within about 0.2 % (in our case). During its way through the quartz capillary along the heated filament the gas beam changes from laminar (hydrodynamic) flow (10-300 Pa) to molecular flow conditions ($\sim 10^{-3}$Pa). In such a set-up decomposition of AsH_3 up to 90 % is achieved.

A different principle is applied in the high pressure effusion source (Fig. 1b) being used by Panish (1980). AsH_3 and PH_3 at a pressure between 0.2 and 2 atm. are injected through alumina tubes with fixed small leaks into the UHV growth chamber. These tubes are mounted within the UHV system in an electrically heated oven, where the hydrids are thermally decomposed effectively by gas phase collisions at temperatures between 900 and 1000°C. On their path through the leak there should be a transition from hydrodynamic to molecular flow. The leak is essentially a free jet. Behind this

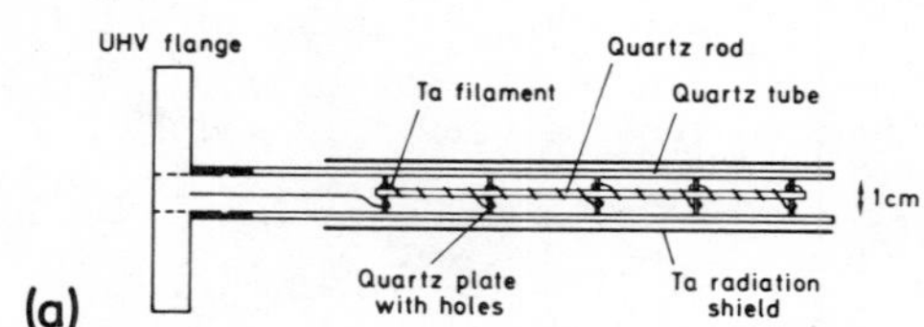

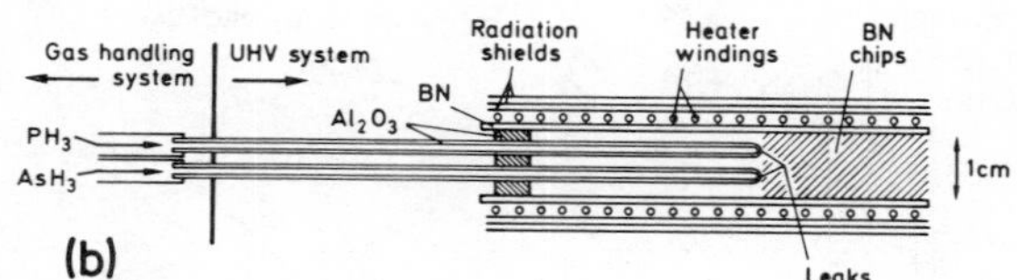

Fig. 1 Low and high pressure gas sources for precracking of the hydrides (AsH_3, PH_3 etc.)

a) Low pressure source with heated Ta filament as used in our group (Pütz et al 1985)

b) High pressure source as used by Panish and Sumski (1984)

leak the decomposition products are injected into a heated low pressure zone where the pressure is in the millitorr range or lower, since this region is directly pumped by the system vacuum. In contrast to the low pressure sources, the flux in this high pressure cell is controlled by pressure variation in the alumina tube with its constant leak. Reproducible flux control requires a constant leak rate.

It seems too early with the present state of knowledge to make any statements about the best final solution to the problem of precracking the hydrides in a MOMBE system for production. Both types of cracking devices seem to be efficient in decomposing the hydrides in principle, whereas the exact ratios for the different As_x species being obtained should be scrutinized more closely.

In contrast to standard MBE, where the UHV system is usually pumped by ion pumps, MOMBE has been performed by means of cryopumps in combination with turbomolecular or diffusion pumps.

3. Growth and Doping of GaAs in MOMBE

The growth of GaAs in the MOMBE system has been studied in most detail so far. TMG and TEG are used as element III sources whereas the element V material is supplied as AsH_3 (Pütz et al 1985) or as an As-alkyl (TEAs etc.) (Tsang 1984). AsH_3 as well as the more stable TEAs have to be precracked in the injection capillary in order to obtain growth of GaAs. Using both TMG and TEG, the grown GaAs layers exhibit p-type background doping due to incorporated C. With TMG the growth rate of GaAs depends both on the As and on the Ga supply as long as the Ga compound is supplied in excess (Pütz et al 1985). Preadsorbed As seems to enhance the adsorption and the decomposition rate of TMG. On the other hand, TEG decomposes without the action of preadsorbed arsenic such that in cases where the Ga supply exceeds the As supply Ga is always deposited on the substrate. Corresponding with their different thermal stability TMG and TEG cause totally different amounts of incorporated carbon, i.e. p-type background doping (Pütz et al 1986). With TMG doping levels above $10^{19}cm^{-3}$ are obtained (Fig. 2), whereas the use of TEG is necessary to produce high quality material with background dopings between 10^{14} and $10^{16}cm^{-3}$ (Fig.2).

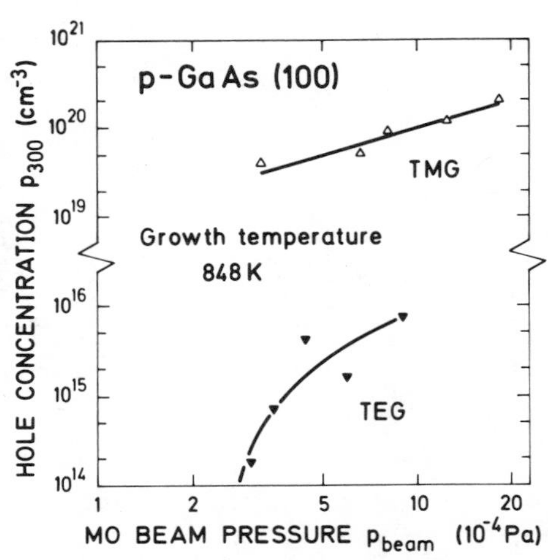

Fig. 2 Effect of metalorganic beam pressure p_{beam} on the background carbon doping, i.e. 300 K hole concentration of GaAs for growth with TMG and TEG. Wafer orientation (100); AsH_3 beam pressure $3x10^{-4}$Pa with maximum cracking efficiency (~85%) (after Pütz et al 1985)

The different amount of carbon incorporation with TMG and TEG is at least partially due to the fact, that TEG can break apart via the so called β-elimination process (Wilkinson et al 1982), where $Ga(C_2H_5)_3$ decomposes stepwise into components $GaH_n(C_2H_5)_{3-n}$ with simultaneous formation of the "inert" C_2H_4 removing carbon from the growing surface. β-elimination is not possible in the TMG decomposition. This might explain the higher amount of carbon incorporation with TMG.

The observation of high carbon incorporation by using TMG in the MOMBE process has led to a new technique of intentional p-type doping of GaAs by carbon (Weyers et al 1986). TEG is used as the basic Ga source and controlled amounts of TMG are added to achieve the desired doping level. By variation of the alkyl beam pressure and the TMG/TEG ratio adjustable hole concentrations between 10^{14} and $10^{21}cm^{-3}$ at 300 K can be produced (Fig.3). The obtained hole mobilities at 300 K are comparable with the best literature values obtained by different growth techniques (Wiley 1975). The high quality of the films is also indicated by the photoluminescence (PL) spectra in Fig. 4 (Weyers et al 1986). The exciton lines between 1.51 and 1.52 eV

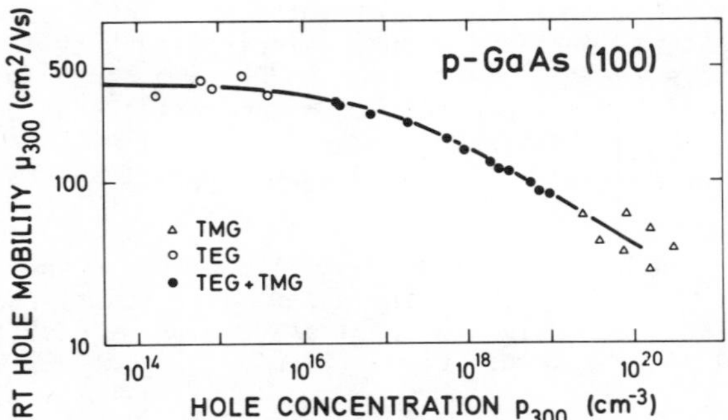

Fig. 3 Room temperature Hall mobilities μ_{300} of intentionally p-doped GaAs layers [orientation (100)] by use of TMG only (▵), TEG only (○), and mixtures of both alkyls (●). Solid line: best mobilities from literature for various growth techniques (Wiley 1975). (after Weyers et al 1986)

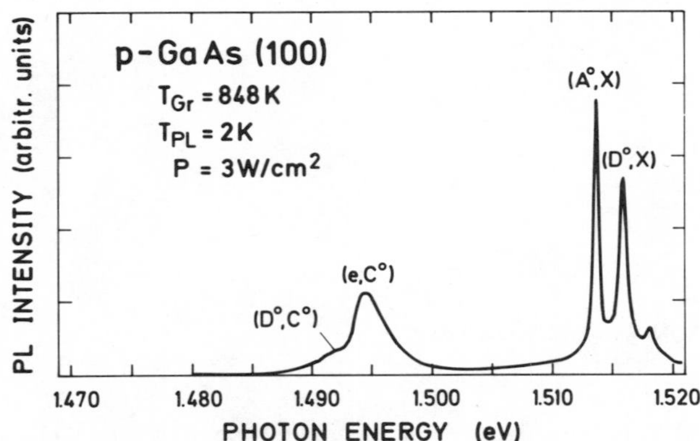

Fig. 4 Photoluminescence (PL) spectrum measured on a lightly p-doped GaAs (100) layer (C-doped: $p \simeq 1x10^{15}cm^{-3}$); growth temperature T_{Gr}=848 K, measurement temperature T_{PL}=2 K, LASER power P=3 W/cm^2 (after Weyers et al 1986)

are extremely sharp as for MBE layers and carbon is detected as a contamination by the characteristic emission near 1.495 eV. The degree of compensation in these MOMBE grown p-type GaAs films is thus comparable with the best MBE samples.

Also, n-type doping of GaAs by Si has been achieved by a gas line source supplying SiH_4 (5 % in H_2) as the dopant (Heinecke 1986a). In principle doping is possible with undecomposed SiH_4, but higher doping levels between 10^{16} and $5\times10^{18}cm^{-3}$ (300 K electron density) can only be achieved at reasonable substrate temperatures by cracking the SiH_4 in the injection capillary at temperatures around 850 K. The Si incorporation, i.e. the doping can be varied by changing the beam pressure or the cracking temperature in the capillary (cracking efficiency). The quality of the obtained n-type layers can be seen from the Hall mobilities measured at 77 and 300 K (Fig. 5). After correction for space-charge layer effects on thin films (1-2 µm in our experiments) the measured room-temperature mobility data of our MOMBE grown layers agree with the best empirical data reported by Stringfellow (1979).

An important advantage of MOMBE over MBE concerns the obtainable surface perfection. Surface defects, in particular oval defects with a typical size in the 10 µm range, are typically produced in MBE with densities higher than 10^3cm^{-2}. Much effort is necessary to reach levels considerably lower than 10^3cm^{-2} even with the best wafer material. This limits severely the applicability of MBE grown layers for the production of integrated circuits.

In contrast, in MOMBE Pütz et al (1985) and Tsang (1985a) could show that the large oval defects of MBE grown layers are essentially absent and that surface defect densities are extremely low. Similar results were obtained recently with TMG/TEG and AsH_3 by Werner et al (1986). On "good" wafer

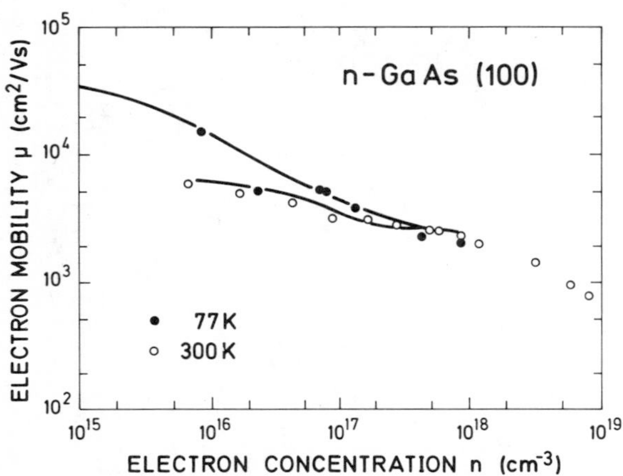

Fig. 5 Electron Hall mobilities µ at 77 K and 300 K of n-doped GaAs(100) films versus electron concentration n; n-doping by use of precracked SiH_4 (5% in H_2). Solid line represents literature values from various techniques (Poth et al 1978) (after Heinecke et al 1986a)

material only some small defects having a typical size of around 1 µm and a characteristic shape as shown in Fig. 6 were observed. Defect densities considerably below 100 cm^{-2} could be achieved. But as in MOCVD (Chang et al 1981) the As/Ga ratio during growth influences the defect density; with increasing As supply more defects are produced. Lower As/Ga ratios, on the other hand, imply higher carbon incorporation (Fig. 2). For certain purposes, therefore, a compromise between tolerable background doping and surface defect density has to be found.

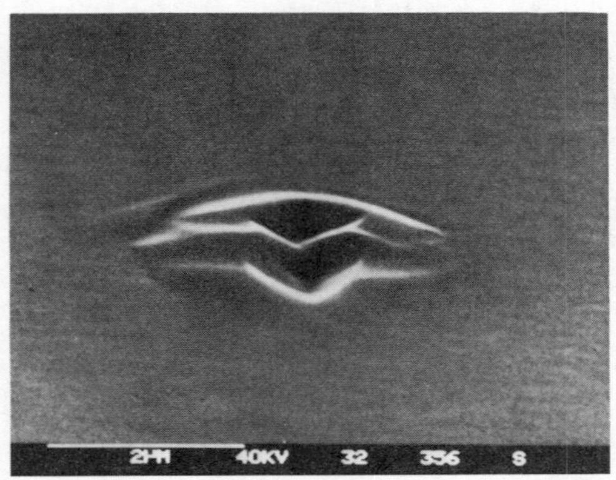

Fig. 6 Scanning electron micrograph of a typical surface defect as occurring on GaAs layers grown in MOMBE by AsH_3 and TMG or TEG. Marker represents 2 µm (after Werner et al 1986)

4. Structures and Devices produced by MOMBE

High pressure gas line sources for AsH_3 and PH_3 in combination with standard effusion cells for Ga and In have already been used in an UHV system by Panish and Tempkin (1984) to grow double heterostructure and separate confinement heterostructure lasers of $Ga_xIn_{1-x}As_{1-y}P_y$ lattice matched to InP.

The complete MOMBE approach with gas line sources for TEG, TMG, AsH_3 and SiH_4 was applied recently by our group (Heinecke et al 1986) to grow modulation-doped multilayer structures (nipi superlattices). The GaAs was grown by means of TEG and AsH_3 as the basic material and p-and n-type doping with a nominal concentration in the $10^{18}cm^{-3}$ range was achieved by additional TMG and SiH_4 (precracked) gas lines. For the example in Fig. 7 the n- and p-type doping period, each 400 Å thick, was repeated 10 times. The corresponding SIMS pattern of Si versus sputter depth of this structure shows a decrease of the Si concentration with a decay length of 5-15 nm with increasing sputtering depth. The background and also the sharpness of the profile is determined partially by the analysis techniques and partially by interface roughness at greater depths. The complementary carbon profile of the p-type regions could not be determined because of experimental limitations of the SIMS equipment. The performance of the nipi structures produced in the described way was tested by photoluminescence (PL) measurements. Due to the spatial modulation of the GaAs band structure in the nipi superlattice (inset in Fig. 7) free electrons and holes are spatially separated. For an opitcal excitation or deexcitation (in PL) an effective gap E_g^{eff}, smaller than in homogeneously doped material, is relevant. Light induced generation of electron hole

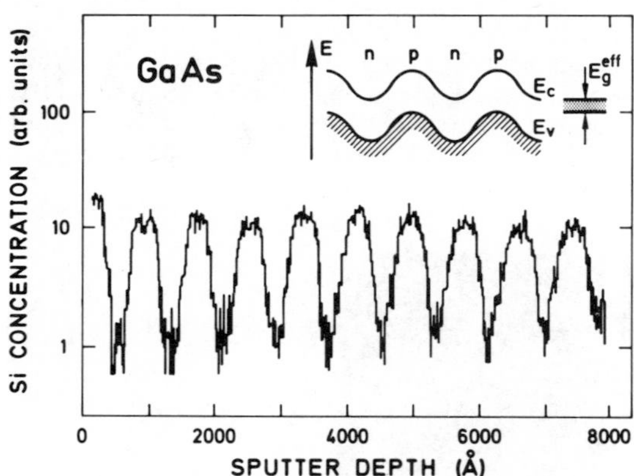

Fig. 7 Spatial variation of Si concentration (SIMS depth profile) of a GaAs(100) nipi doping superlattice with 10 periods (800 Å thick); p-doping by C (TMG) and n-doping by Si (SiH_4) with nominal concentrations of about $10^{18}cm^{-3}$. Inset: qualitative band structure of a nipi superlattice (after Heinecke et al 1986a)

pairs decreases the space charge and thus smoothes out the band modulation. The effective gap increases. This is exactly found in the PL spectra of Fig. 8 (Heinecke et al 1986a). For higher excitation densities the luminescence band shifts from a value of 1.33 eV towards the GaAs band gap (1.519 eV at 2 K). The results obtained with MOMBE grown layers agree well with those found on nipi structures produced in standard MBE (Döhler et al 1981).

The practibility of MOMBE for the growth of LASER structures was recently demonstrated by Tsang (1986). Using gas line sources for TEG (with hydrogen as carrier gas), TMAl and AsH_3 (low pressure cracker), he grew a double

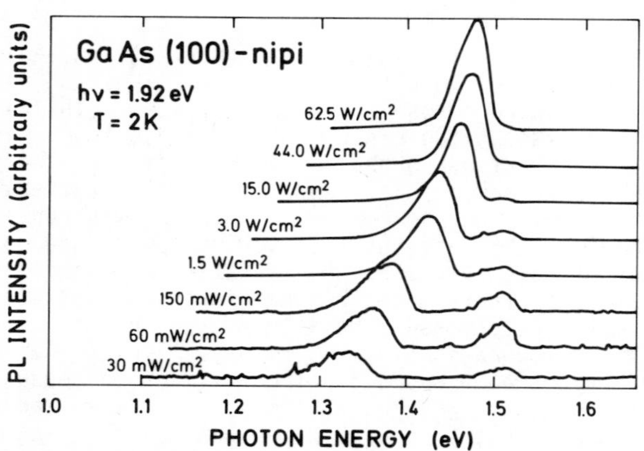

Fig. 8 Photoluminescense (PL) spectra of the GaAs(100) nipi superlattice of Fig. 7, measured at a temperature of 2 K with photon energy of 1.92 eV and various LASER excitation powers. (after Heinecke et al 1986a)

heterostructure (DH) LASER consisting of five layers: ~1.0 μm n-GaAs (~ $5x10^{18}cm^{-3}$, buffer layer), ~2.5 μm N-$Al_{0.5}Ga_{0.5}As$ (~ $5x10^{17}cm^{-3}$, confinement layer) undoped GaAs active layer, ~2.0 μm P-$Al_{0.5}Ga_{0.5}As$ (~ $1x10^{18}cm^{-3}$, confinement layer), and 0.3 μm p^+-GaAs (~ $1x10^{19}cm^{-3}$ cap layer). For p- and n-type doping the conventional MBE approach was used, ie evaporation of Be and Sn from effusion cells, respectively. The current threshold density J_{TH} was evaluated on broad area LASERS fabricated from each wafer. The area of the diodes was 375 μm x 200 μm with two cleaved mirrors and two scribed sidewalls. The current pulses were ~1 μs pulse per second. A very uniform lasing action was observed across the entire facet simultaneously at threshold, indicating excellent uniformity of the material quality. A comparison of averaged threshold current densities J_{TH} as a function of active layer thickness for the best MOMBE grown and the best MBE grown DH LASERS is given in Fig. 9. For active layers thinner than 1000 Å J_{TH} reaches values lower than 500 A/cm^2. This is comparable to the best results from MOCVD (Dupuis and Dapkus 1978) and MBE. The hetero-interfaces thus should have an excellent electrical and optical quality. This is also revealed from PL data which have been obtained on single and multiple $Al_{0.5}Ga_{0.5}As$/GaAs quantum wells grown by the same technique (Tsang and Miller 1986).

Using a similar experimental method (TEG, TMIn, AsH_3, PH_3 and elemental Be and Sn as dopants) Tsang and Campbell (1986) produced mesa type InGaAs/InP p-i-n photodiodes with a very low dark current (less than 1 nA at -10 V bias) and a remarkably good quantum efficiency of 70 %. The device characteristics suggest the excellent quality of the heterojunction interfaces in this case also.

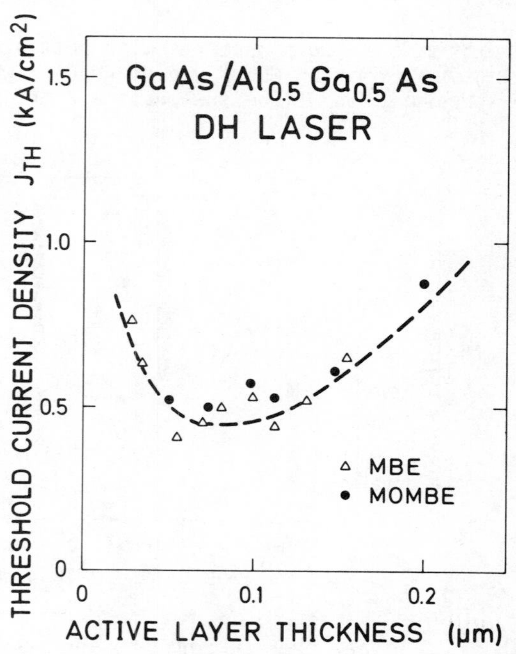

Fig. 9 Threshold current densities J_{TH} versus active layer thickness for different DH LASERS grown in the system GaAs/$Al_{0.5}Ga_{0.5}As$ by means of MBE and MOMBE (or chemical beam epitaxy). Each data point represents an averaged J_{TH} for each wafer. The cavity length used was 375 μm in all cases (after Tsang 1986)

5. Selective Growth

For production of optoelectronic devices and in particular for their integration into circuits, lateral structuring during the epitaxial process itself can lower the number of process steps. The possibility of selective growth in certain spatial areas of the wafer and suppression of growth in others is therefore highly desirable. One approach both in low pressure MOCVD and in MOMBE uses mechanical masks to shield part of the wafer against deposition (Bedair et al 1986, Asai and Ando 1985, Tsang 1985b). This method of course is limited in its application because of poorly defined edges and finite beam divergences. Much more attractive and versatile is the technique where glassy protective layers (SiO_2 or SiN_x) are pyrolytically deposited on the wafer and structured by photolithographic processes. Subsequent wet chemical etching or dry reactive ion etching (RIE) removes limited areas of the masking layer where the clean wafer surface is then exposed to the gaseous source material. This masking layer approach to selective growth has been investigated both in MOCVD and in MOMBE for the growth of GaAs (Balk 1985, Heinecke et al 1986b, Lüth 1986). As a general conclusion one can say that selective growth in the strict sense, i.e. crystalline growth in the open areas and no (polycrystalline) deposition on the mask, is favoured at lower pressures (Fig.10). The use of TMG enables selective growth in a much broader temperature range (100 to 150 K) than does TEG. On the other hand, in MOMBE the use of TEG is required to reach low background p-type doping because of incorporated carbon. Thus, in MOMBE essentially a temperature range between 870 and 900 K can be used for selective growth of GaAs with TEG and AsH_3 (Fig. 11). Below 870 K selective growth breaks down and polycrystalline deposit occurs on the mask, similar to MBE.

If extremely sharp structures with best electrical properties are required, selective growth by MOMBE yields results which are superior to those of low pressure MOCVD. The sharpness and the edge definition obtained in

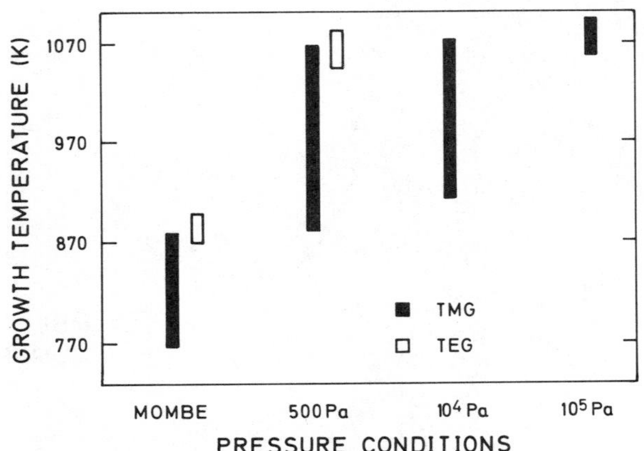

Fig. 10 Overview over temperature and pressure conditions, where selective growth by means of structured masking layers of SiO_2 is possible in MOMBE (typical beam pressure $\sim 10^{-3}$Pa) and in low pressure MOCVD (pressures between 500 Pa and 10^5Pa)

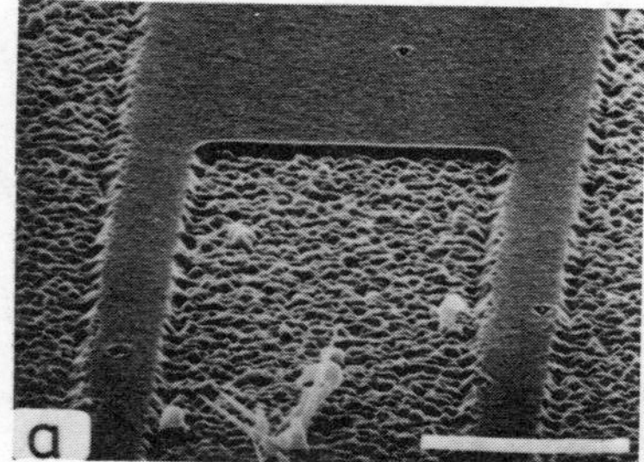

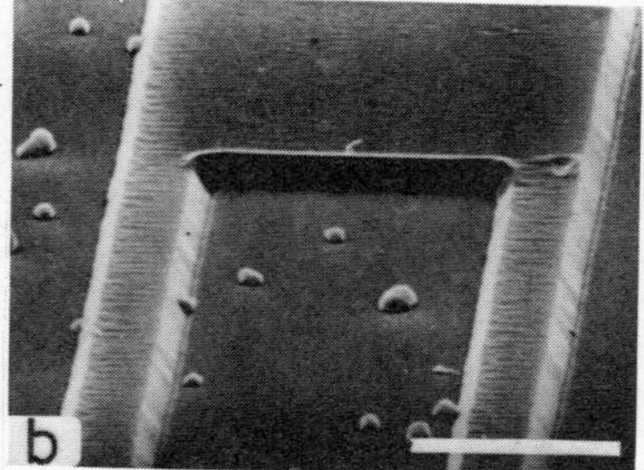

Fig. 11 Scanning electron micrographs of GaAs patterns grown selectively in MOMBE with TEG and AsH_3 at various substrate temperatures: (a) 850 K, (b) 870 K, (c) 890 K. SiO_2 masks were prepared by photolithography and wet chemical etching. (a) demonstrates the break down of selectivity, whereas in (b) single nuclei are found on the SiO_2 mask. Markers represent 10 µm (after Heinecke et al 1986b)

MOMBE is demonstrated in Fig. 12. RIE in contrast to chemical etching yields a much sharper definition of the mask edges (angles around 90°) and thus also of the grown GaAs pattern in between. Overgrowth at the edges of the mask is not observed. GaAs structures even in the submicron range can be produced by this technique (Fig. 12b).

The general possibility of selective growth is explained by a different adsorption and/or nucleation behaviour of the source material on the GaAs substrate and on the mask material. In standard MBE, e.g. Ga excess is necessary for epitaxial growth, but the sticking coefficient of elemental Ga on a SiO_2 mask is high enough at growth temperatures, such that nucleation can occur and therefore also polycrystalline deposition. Selective growth is not possible. On the other hand, in MOMBE growth of GaAs requires sticking and decomposition of the metalorganic compound, TMG or TEG. The decomposition reaction of the alkyl obviously depends on the surface conditions of the GaAs and the SiO_2 mask material, respectively. Since increased As supply under growth conditions with higher As/Ga ratio leads to a break down of selectivity in low pressure MOCVD (500 Pa) and plasma

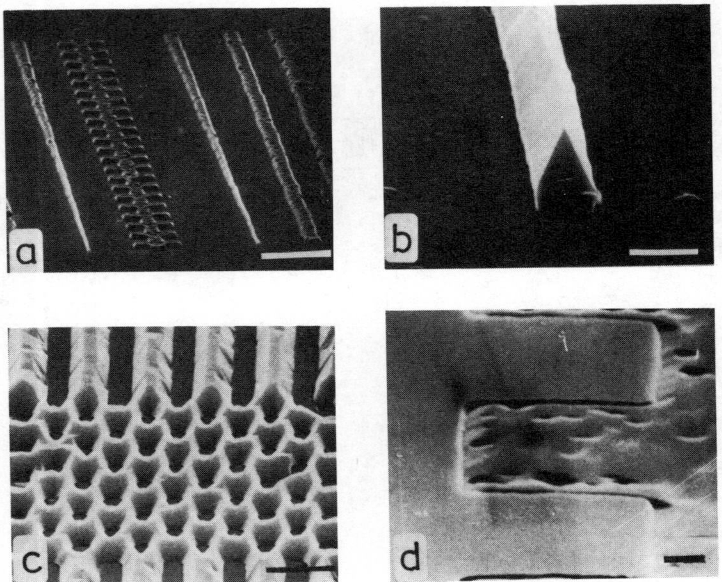

Fig. 12 Scanning electron micrographs of GaAs patterns grown selectively by means of SiO_2 masks (as in Fig. 11) in MOMBE with TEG and AsH_3 at 890 K substrate temperature on GaAs(100). SiO_2 masks were prepared by photolithography and reactive ion etching.
a) overview of structures with various dimensions; the GaAs film grown 1.3 µm thick, marker 20 µm
b) magnified section of submicron stripe end from (a), marker 1 µm
c) honey-comb structure with GaAs film of 1.3 µm thickness, marker 4 µm
d) GaAs film equally thick as SiO_2 mask (0.2 µm), marker 2 µm (after Heinecke et al 1986b)

MOCVD (Heinecke et al 1986b), it is suggested that the presence of some adsorbed As species on the surface is necessary for growth. Selective growth in MOMBE is therefore likely due to the different sticking coefficient for arsenic on the GaAs and on the mask surface. Elemental arsenic or an arsenic complex probably enhances the decomposition of the alkyl and growth is initiated.

6. Conclusion

The present review shows that gas source MBE, in particular MOMBE permits the epitaxial growth of high quality GaAs, InP as well as of ternary and quaternary compounds e.g. AlGaAs and GaInAsP. As was demonstrated for GaAs, both growth and doping can be achieved "in one package" by gas line sources only. The first more complex MOMBE grown structures such as nipi superlattices, DH LASERS and p-i-n diodes show excellent performance characteristics. In general, the MOMBE work is in a comparatively early stage ; to prove its total capability will certainly need a lot more investigation . But it can already now be seen that the quality standards concerning electrical and optical properties of the grown films reach those of the best MBE grown layers. In addition, the problem of surface defects might better be solved in MOMBE than in MBE. With respect to production, the problem of upscaling, i.e. of achieving higher throughput numbers of wafers, might be handled better in MOMBE. The technique also has the potential to be applied to a wider field of applications: Epitaxial metal films like e.g. Fe on GaAs have already been deposited recently by means of an $Fe(CO)_5$ gas source (Kaplan 1983).

Acknowledgement

I would like to thank P. Balk, H. Heinecke, N. Pütz, M. Weyers and K. Werner for their pleasant cooperation in this field of MOMBE and in particular P. Balk and H. Heinecke for critical reading of the manuscript.

References

Asai H and Ando S (1985) J. Electrochem. Soc. 132 2445
Balk P and Heinecke H (1985) in "Physical Problems in Microelectronics" ed. Kassabov J (World Scientific Publ. Co., Singapore) p 190
Bedair S M, Tischler M A and Katsuyama T (1986) Appl. Phys. Lett. 48 30
Calawa A R (1979) Appl. Phys. Lett. 33 1020
Calawa A R (1981) Appl. Phys. Lett. 38, 701
Chang C Y, Su Y K, Lee M K, Chen L G and Houng M P (1981) J. Crystal Growth 55 24
Chow R and Chai Y G (1983) J. Vac. Sci. Technol. A 1 49
Döhler G H, Künzel H, Olego D, Ploog K, Ruden P, Stolz H J and Abstreiter G (1981) Phys. Rev. Lett. 47 864
Dupuis R D and Dapkus P D (1978) Appl. Phys. Lett. 32, 473
Heinecke H, Werner K, Weyers M, Lüth H and Balk P (1986a) J. Crystal Growth, to be published
Heinecke H, Brauers A, Grafahrend F, Plass C, Pütz N, Werner K, Weyers M, Lüth H and Balk P (1986b) J. Crystal Growth, in press
Huet D and Lambert M (1986) J. Electronic Mat. 15 37
Kaplan R (1983) J. Vac. Sci. Technol. A 1 551
Kawaguchi Y, Asahi H and Nagai H (1984) Jap. J. Appl. Phys. 23 L737
Kawaguchi Y, Asahi H and Nagai H (1985) Jap. J. Appl. Phys. 24 L221
Kawaguchi Y, Asahi H and Nagai H (1986) Inst. Phys. Conf. Ser. No 79 79
Kondo K, Ishikawa H, Sasa S, Sugiyama Y and Hiyamizu S (1986) Jap. J. Appl. Phys. 25 L 52
Lüth H (1986) Proceedings of the European MRS conference, Strasbourg 1986, to be published
Nishizawa J and Abe H (1985) J. Electrochem. Soc. 132 1197
Panish M B (1980) J. Electrochem. Soc. 127 2730
Panish M B and Sumski S (1984) J. Appl. Phys. 55 3571
Panish M B and Tempkin H (1984) Appl. Phys. Lett. 44 785
Poth H, Bruch H, Heyen M and Balk P (1978) J. Appl. Phys. 49 285
Pütz N, Veuhoff E, Heinecke H, Heyen M, Lüth H and Balk P (1985) J. Vac. Sci. Technol. B 3 671
Pütz N, Heinecke H, Heyen M, Balk P, Weyers M and Lüth H (1986), J. Crystal Growth 74, 292
Stringfellow G B (1979) J. Appl. Phys. 50 4178
Tempkin H, Panish M B, Petroff P M, Hamm R A, Vandenberg J M and Sumski S (1985) Appl. Phys. Lett. 47 394
Tokumitsu E, Kudou Y, Konagai M and Takahashi K (1984), J. Appl. Phys. 55 3163
Tokumitsu E, Kudou Y, Konagai M and Takahashi K (1985) Jap. J. Appl. Phys. 24 1189
Tsang W T (1984) Appl. Phys. Lett. 45 1234

Tsang W T (1985a) Appl. Phys. Lett. 46 1086
Tsang W T (1985b) Appl. Phys. Lett. 46 742
Tsang W T (1986) Appl. Phys. Lett. 48 511
Tsang W T and Miller R C (1986) Appl. Phys. Lett. 48 1288
Tsang W T and Campbell J C (1986) Appl. Phys. Lett. 48 1416
Veuhoff E, Pletschen W, Balk P and Lüth H (1981) J. Crystal Growth 55 30
Vodjdani N, Lemarchand A and Paradon H (1982) J. Phys. Colloq. C5 43 339
Werner K, Heinecke H, Weyers M, Lüth H and Balk P (1986) J. Crystal Growth, to be published
Weyers M, Pütz N, Heinecke H, Heyen M, Lüth H and Balk P (1986) J. Electronic Mat. 15 57
Wiley J D (1975) Semiconductors and semimetals, Vol. 10, edited by Willardson R K and Beer A C (1975) Academic Press, New York p.154
Wilkinson G, Stone G A and Abel E W, Editors (1982), "Comprehensive Organometallic Chemistry", Vol. 1, Pergamon Press, Oxford

Inst. Phys. Conf. Ser. No. 82
Paper presented at ESSDERC 1986, Cambridge 8–11 Sept. 1986

2D Device modelling

P J Mole

GEC Research Ltd, East Lane, Wembley, Middlesex HA9 7PP.

ABSTRACT

In this paper a description of both the numerical techniques and the physical models used in present day semiconductor device simulators is given. These are presented in a manner which should enable users to understand the strengths and the limitations of these tools. In particular, the model for the carrier mobility at a silicon surface is presented in detail. The interrelationship of process, device and circuit simulators is also discussed.

1. INTRODUCTION

The development of advanced silicon devices for both the power and integrated circuit industries is, technologically, very challenging. The complexity of modern manufacturing processes combined with the expense of equipment used during fabrication indicate that if such a development is undertaken empirically it will be both slow and very expensive. Device simulation has become popular because it has helped the developers of new technologies to understand operation of their devices and quantitatively predict the effect of process changes, thereby reducing technology development times and costs. To allow for the spreading of electric field lines about features in the devices, multi-dimensional device simulation is required. Although there is a requirement for simulation in 1,2 and 3 dimensions, 3 dimensional simulation is, as yet, uncommon because it places a high demand on computer resource. 2 dimensional simulation, however, is widely used, with a variety of packages now available throughout the world.

It is the objective of this paper to give an overview of both the physical models and the numerical techniques used within current 2 dimensional simulators in order to aid users of the packages in interpreting the behaviour of their simulation tools. The ideas, although addressed at 2 dimensional simulators, can be applied equally to 1 (or 3) dimensional simulators. To start the paper, an example of what the device designer can learn from simulation is presented.

2. APPLICATION OF DEVICE SIMULATION

As an example of the application and interpretation of numerical simulations, figure 1 compares the electron distributions simulated within a long (3 μm) channel length N-channel MOSFET and a short (1.2μm) device. Both devices have a substrate doping level of 5.10^{15} cm^{-3}, and a gate oxide thickness of 30 nm. The bias applied to the gate is zero volts, and, as a consequence, the device is weakly turned off. Whilst the longer device shows negligible current flow with a 10V drain bias, the shorter device passes a significant current.

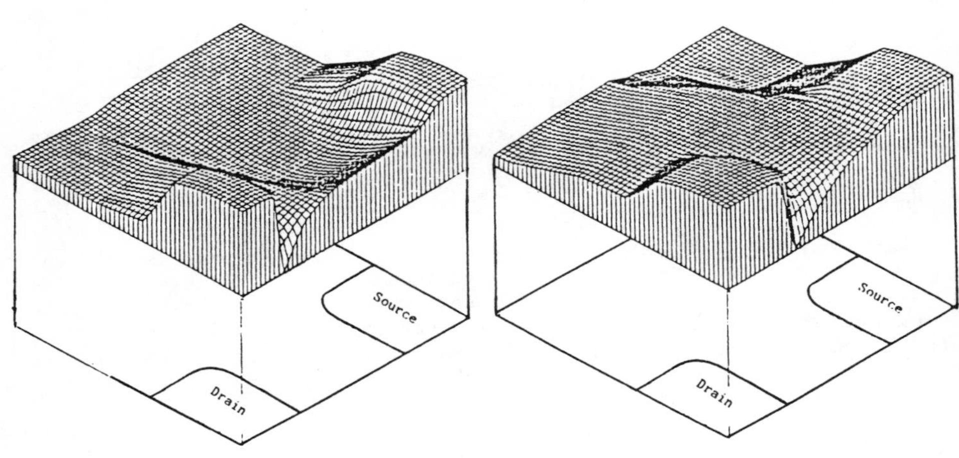

3.0 µm long device 1.2 µm long device

Figure 1. Representation of the distribution of electrons in an N-channel MOSFET (Drain bias = 10V, Gate bias = 0V)

In the case of the longer device, the source and drain regions can be clearly identified by their high electron density. A ridge of high electron density can be seen along the surface (next to the oxide) and this is the weak remains of the channel and is not large enough to support a significant current. Close to the drain, the surface carrier density falls off rapidly; this is akin to channel pinchoff as the device current saturates.

In the case of the smaller device, however, a plateau of high carrier density has been introduced which extends from the source to the drain. This plateau has been introduced by the merging of the drain and source depletions and is reponsible for the high 'punchthrough' current observed. This current is no longer confined to the surface, the simulation indicating that the plateau extends out to the source and drain junction bottoms. Thus device simulation has shown that it is not possible to make 1.2µm FETs which operate at 10V from such a simple process, because of punchthrough. But let us suppose that it was necessary to improve the performance of a 1.2µm FET by making adjustments to the process without affecting the FET's threshold voltage. What is the best approach? Clearly to prevent punchthrough the source and drain lateral depletion widths must be narrowed; but the doping cannot simply be increased to effect this without causing the threshold voltage to increase.

Two approaches have been investigated, by way of example:

a) The peak doping has been increased close to the source and drain junction bottoms where punchthrough dominates. The substrate doping was reduced to maintain a constant long channel threshold (using 1 dimensional simulation).

b) The substrate doping was increased to $2.10^{16}cm^{-3}$ and a shallow N type implant was used to adjust the threshold voltage.

Figure 2 compares the doping profiles used in the simulations. Figure 3 compares the drain current/drain voltage behaviour for these approaches with the uniformly doped starting process.

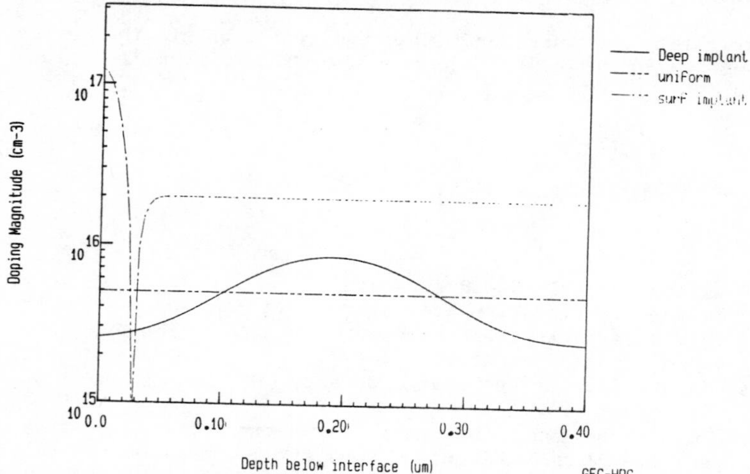

Figure 2. Channel doping profiles investigated

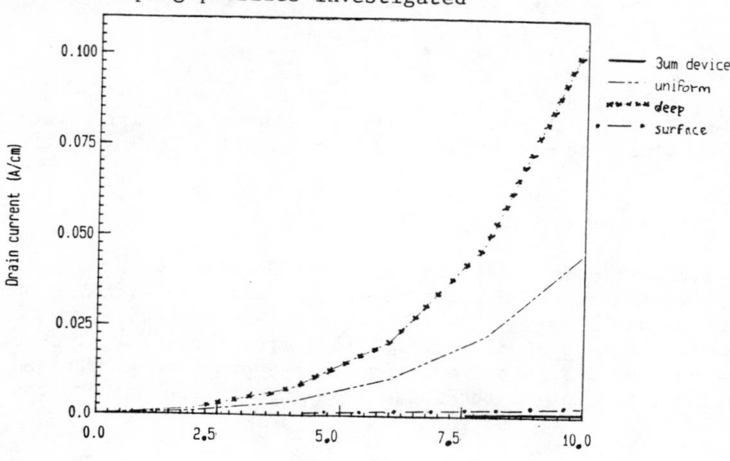

Figure 3. Device drain current with gate turned off for selected channel dopings

Approach a) can be seen, in fact, to make matters worse whilst b) effects a significant reduction in current but the leakage is still noticeable. Investigation of the current flow in a) indicates that the reduction of substrate doping has given rise to increased punchthrough deep in the substrate, behind the doping peak.

Thus with the help of device simulation it is straightforward to reduce the number of process variations that need be tried experimentally by eliminating ineffective variants. Thus processing time and costs can be reduced.

3. THE PHYSICS BEHIND THE DEVICE SIMULATOR

3.1 Poisson's equation

The distribution of charge and electric field within the device must be consistent with Maxwell's equations. Thus, from the two relevant equations:

$$\underline{\nabla} \times \underline{E} = -\partial \underline{B}/\partial t = -\partial(\underline{\nabla} \times \underline{A})/\partial t \qquad 1$$

$$\underline{\nabla D} = \rho \qquad 2$$

where: $\underline{E}$ is the electric field strength
$\underline{D}$ is the electric displacement = $\varepsilon\varepsilon_0\underline{E}$
$\underline{B}$ is the magnetic field strength
$\underline{A}$ is the vector potential
ρ is the local space charge density

It should be noted that the cubic symmetry of the lattice must be acknowledged when the dielectric constant is assumed to be a scalar quantity. Following the classic lines of electromagnetic field theory equation 1 can now be used to define the electrostatic potential (Ψ) as:

$$\underline{\nabla} \Psi = -(\underline{E} + \partial \underline{A}/\partial t) \qquad 3$$

This determines Ψ to within an additive constant whilst ensuring 1 is satisfied. Using 3 to substitute for $\underline{D}$ in equation 2:

$$\underline{\nabla}(\varepsilon . \varepsilon_0(\underline{\nabla\Psi} + \partial \underline{A}/\partial t)) = -\rho \qquad 4$$

The time derivative of $\underline{A}$ can be neglected provided we are not concerned with events which happen on the time scale associated with the propagation of electromagnetic waves in the device. Furthermore, the local space charge density is made up of contributions from the local electron density (-qn), the hole density (qp) and the charge provide by ionised impurity atoms (qN) which can be obtained from information about the device's processing.

$$\underline{\nabla}(\varepsilon . \varepsilon_0 \underline{\nabla\Psi}) = -q(p-n+N) \qquad 5$$

This is the familiar form of Poisson's equation used in all device simulation. It should be noted that it is not applicable if wave propagation affects are significant, and that the semiconductor must have sufficient symmetry to have a scalar dielectric constant.

3.2 The current conservation equations

When a current is flowing within a device, it is not adequate to consider

Poisson's equation alone; the flow of both electrons and holes must satisfy conservation criteria. In this section, only the equations for holes are considered, however an exact analogue can be derived for the electron equation.

Firstly, the conservation of number density (p) must be considered; this requires:

$$\partial p/\partial t = -\underline{\nabla}\ \underline{J}p/q + G \qquad 6$$

where Jp, is the local current density and G is the carrier pair generation rate.

The generation rate G must represent all significant generation and recombination mechanisms active in the device. This representation is restricted, for practical reasons, to functions of the local potential, electric field and carrier density. Such a limitation is not of concern for Shockley, Read (1952), Hall (1952) trap-assisted recombination or for Auger recombination, (Adler et al 1981). However, for impact ionisation, Kunhert et al (1985) have pointed out that the ionisation rate is dependent on the electric field that a carrier has seen since it was last scattered, thus the local electric field dependence given by Chynoweth (1957) should be replaced by a non-local dependence if the electric field varies rapidly over distances comparable to the inelastic scattering length for the carriers. The generation rate used within device simulators is of the form:

$$G = \frac{n_i^2 - pn}{\tau_p(n+n_i)+\tau_n(p+n_i)} - (C_n\, n^2\, p + C_p\, p^2\, n)\ [\text{SRH + Auger}]$$

$$+\ |J_n|\ \exp-(B_n/E_n) + Ap|Jp|\ \exp-(Bp/Ep)\ [\text{Impact}] \qquad 7$$

where Ep is the component of electric field in the direction of the local hole current flow (and similarly for E_n) and A,B,C and τ are parameters which characterise the models.

Of the parameters which control generation it should be noted that τ is an extrinsic property of silicon and will depend on the silicon source and subsequent processing. There is considerable debate in the literature (e.g. Adler et al (1981))as to whether C is dominated by extrinsic effects, the argument being hindered by the practical difficulty in measuring C in isolation from other effects. A and B are believed to be intrinsic properties of the film.

Secondly, the average carrier momentum must be conserved. Thus:

$$\underline{J}_p/(q.\mu) = p\ \underline{E} - \underline{\nabla}\ [p.kT_p] \qquad 8$$

where T_p represents the temperature of the carriers. To see that this equation is equivalent to that of momentum conservation, note that the current $\underline{J}p$ is proportional to the mean carrier drift velocity and hence the net momentum. This momentum decays with a rate characterised by the momentum relaxation time which governs the carrier mobility (μ). The first term on the right hand side represents the rate at which the carriers acquire momentum from the electric field, whilst the second term represents the rate at which momentum 'diffuses' from the neighbouring regions because either there is a greater carrier density there, or the carriers have a greater individual momentum as they are 'hotter'. Thus, in the simple case, where the carriers are in thermal equilibrium with an isothermal substrate, the second term reduces simply to a diffusion term, and the Einstein relationship defining the relationship between the diffusivity and the mobility drops out implicitly. Blotekjaer (1970) details the argument deriving this equation from the Boltzmann transport equations.

Thus equation 8 demands that the rate at which carriers lose momentum by scattering must exactly balance the rate at which the local population of carriers gain momentum from the electric field and by diffusion. It is more commonly rearranged as:

$$\underline{J_p}/q = \mu\ p\ \underline{E} - \mu kT\ \underline{\nabla}\ p \qquad \text{8a}$$

for use in device simulators where carrier heating is ignored.

Equations 6 and 8 combine together (8b) to provide the constraint on current conservation which must be solved consistently with the Poisson equation.

$$\partial p/\partial t = -\ \underline{\nabla}(\mu\ p\ \underline{E} - \mu\ kT\ \underline{\nabla}p) + G \qquad \text{8b}$$

The mobility must be suitably modelled as a function of local variables in a manner similar to the recombination. This is covered in more detail in section 7.

3.3 The energy conservation equations

Further to the conservation of carrier density and momentum, carrier energy must also be conserved. (In the limit there must be a complete balance of the distribution function over all momenta and space, although current practice is only now finding it important to consider energy balance as well as density and momentum.) The inclusion of the energy conservation equation is currently growing in interest (Selberherr (1986)) and it is useful here to see what the problem involves.

Firstly, a mean local carrier energy E can be defined together with its equilibrium value E_o. There will be a flux of energy $\underline{\eta}$ through the device. The carriers in a given locality will gain energy from the electric field, from the diffusion of energetic carriers and from any release of energy (U_t) during recombination (e.g. Auger recombination). In the steady state a balance must be achieved thus:

$$(E-E_o)/\tau_e = \underline{Jp}.\underline{E} + U_t - \underline{\nabla\eta} \qquad 9$$

In this equation, the relaxation time τ_e is the energy relaxation time and will not usually be the same as the momentum relaxation time which dominates the mobility. The greatest difficulty comes in obtaining a suitable expression for the energy flux $\underline{\eta}$ (Hansch and Miura-Mattausch (1986)). If a suitable expression for $\underline{\eta}$ can be derived then the carrier energy (and hence temperature) will be influenced by its surroundings. This will enable impact ionisation to be calculated as a function of carrier temperature and hence overcome the arguments associated with it being dependent only on local electric field. Moreover, it would allow effects due to velocity overshoot to be calculated along the lines of Baccarani and Wordeman (1985). However, in the meantime some of the effects of carrier heating have been included in MINIMOS 3 by ignoring the last two terms in equation 9 and hence assuming that the carrier energy can be directly related to the local electric field. If the carrier 'temperature' calculated thus is substituted in equation 8 the effects of enhanced momentum diffusion can be included within the simulation. For MOSFETS this means that effects due to the broadening of the inversion layer as the carriers heat up can be included in the simulator.

3.4 Modelling of contacts to the silicon

The discussion so far has been centered on the modelling of the bulk of the device. However, the injection of current at the contacts can be important to the operation of the device. By far the most common boundary condition at a contact is that of the ohmic contact. In this model the carriers on each side of the contact are assumed to be in perfect diffusive contact. Furthermore, on the silicon side of the contact the electrons and holes are assumed to be in thermal equilibrium: this corresponds to demanding an infinite surface recombination velocity at the contact. The equations which model this behaviour are:

$$pn = n_i^2$$

$$Vext = \Psi + (kT/q)\ln(p/n_i) = \Psi - (kT/q)\ln(n/n_i) \qquad 10$$

Where Vext is the applied external bias voltage.

For MOSFETS this type of boundary condition serves the simulator well, but in any device with bipolar action it must be ensured that either the contact is considerably further from an active depletion region than one diffusion length, or that an infinite surface recombination rate is believed to be a good model for the actual contact present in the device.

A simple modification to this model is to include a series resistance. This does not in any way change the physical mechanisms present at the contact interface itself.

The second type of contact that can be included in a device simulator is the Shottky contact, In this model a surface recombination velocity v_r, which is a property of the metal and semiconductor forming the contact, is defined. This governs the rate at whch carriers pass over a potential barrier, the height of which (Ψ_{barr}) is a property of the contact materials. Thus boundary conditions at the surface are of the form:

$$J_{norm} = q\,(n-n_o)\,v_r$$

$$\Psi = Vext - \Psi_{barr} \qquad 11$$

These are adequate to describe the contact when v_r, n_o and Ψ_{barr} are prescribed (Sze (1981)). The potential will vary rapidly in a depletion region induced in the silicon by the contact. The space of this depletion region will be calculated by the simulator together with any currents generated there by traps etc.

Although this provides a flexible description of a contact it is not currently a widely used contact model. Maybe the latest generation of bipolar devices with polysilicon emitters will stimulate the use of such contacts.

4 THE DEVICE SIMULATION MESH

The physical models used to simulate a device were presented in the preceeding section in the form of a coupled set of partial differential equations. At a minimum, this set of equations consist of Poisson's equation (5), the hole current continuity equation (8b) and its electron counterpart. It is not possible to describe, in general, a continuous function that will satisfy these equations simultaneously at every point in the device. This is overcome by inserting a finite number of points within the device and at each point it is assumed that a value of the potential, electron and hole concentration will be calculated. A suitably chosen interpolation will then allow these values to be extended to any other point within the device; if the number and position of the points are well chosen this should not introduce a significant error. The potentials at the points are calculated to satisfy the partial differential equations, not exactly at every point, but in an average sense in the neighbourhood of each point. This results in a function for each mesh point which must be equated to zero by a suitable choice of variables. This procedure is known as discretisation; in this section the aim is to describe how the points are selected and to indicate how the error introduced by the discretisation can be assessed.

There are two approaches to the generation of the points. Firstly, there is the finite difference approach which is illustrated in Figure 4a. In this approach a Cartesian grid of mesh lines is superposed on the problem, each mesh line starts at one edge of the problem and passes to the opposite one. If there is a region of the device which requires a fine mesh to resolve the detail, then that fine mesh must propogate out to the edges. This type of mesh is very well suited for rectangular structures, it is easy to generate and results in very simple data structures in the simulation software. However, it cannot cope with awkwardly shaped

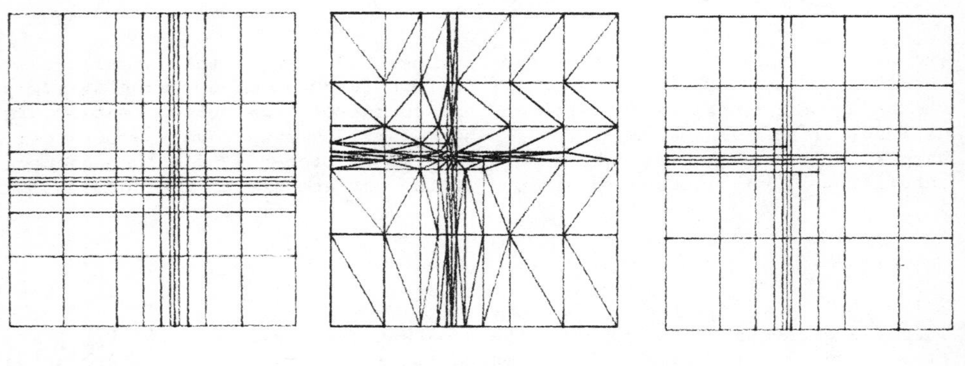

a/ Finite Difference

b/ Finite element

c/ Finite boxes

Figure 4. Types of mesh employed by device simulators

features in the device geometry and it does demand an unnecessarily high number of points in the problem.

In contrast the finite element approach breaks the device down into a finite number of small polygons which pack the device. The most commonly used element shape is the triangle (Figure 4b). Clearly with the triangle, very general geometries can be described by the mesh; moreover the density of mesh points can be graded to give a high density just where the device is interesting. However such a mesh is difficult to construct and leads to more complex data structures within the simulation software.

An intermediate mesh type, finite boxes, has been described which attempts to capture the advantages of both methods. In this case (Figure 4c) any mesh line can end part way through the device. A special strategy (Franz et al (1983)) is then adopted to solve the equations in the neighbourhood of the termination. This approach is very similar to that adopted by many finite element mesh generators which keep the majority of the points on a locally regular grid.

Although here the distinction between finite element and finite difference has been made in terms of the description of the mesh, strictly speaking the method of ensuring the partial differential equations (PDE s) are satisfied locally is also different between the two approaches. In practice all device simulators, regardless of mesh type, use a finite difference approach to converting the PDEs to a set of simultaneous equations.

The method of generating the mesh is, in practice, a very important part of the simulator. Many device simulators of the finite element type rely on a semi-automatic mesh generation where the user provides the outline of the device and indicates where he believes a fine mesh is needed. The mesh generator then divides the domain into elements according to its instructions. This process is completely divorced from the solution and relies on the users 'feel' for the problem (and a posteriori checking) to ensure an appropriate mesh.

A second approach, the fully automatic or adaptive mesh scheme, requires the user to provide a description of the boundary of the device. The simulator then generates its own mesh, usually guided by the doping profile for the device; it then adapts the mesh whilst it is solving the equations in an attempt to minimise the error. This procedure works well for simple finite difference meshes (c.f MINIMOS) but the mesh refinement for finite element meshes is more difficult and as yet is not widely used.

At this point, it is worth asking what criteria are used to determine whether the mesh is adequately accurate. Armed with this information a user can judge whether he is satisfied with the solution obtained or whether he requires more demanding criteria. The criteria used are most commonly based on local variation of the variables (potential and carrier densities) and the doping. Thus a mesh will be refined if the doping varies by more than a factor r or the potential charges by more than δV (usually about 10 thermal volts) between neighbouring nodes. It should be noted that although these are physically reasonable requirements, they cannot be directly related to the numerical errors in the equations solved.

If the user is in doubt as to whether the mesh used is adequate for his problem, then an empirical approach is best to gain confidence in the accuracy of the simulator. If semi-automatic mesh generation is in use, the resolution of a typical problem on a different mesh will give an indication of the error introduced by the mesh. If an adaptive scheme is used, then the relevant criteria must be changed; this may sometimes require alteration of the source code. This should be treated as an exercise to establish confidence for an example of the user's problem.

Before leaving this section, a few words about the interpolation of variables in the problem are in order. The mesh must be sufficiently fine to ensure accurate interpolation of the variables between mesh points: however in a real semiconductor device the carrier density can easily change by ten orders of magnitude in less than 1μm. At first sight adequate meshing of such a variation appears to be a hopeless task and it might be believed that very dense meshes are required. However, based on an observation by Scharfetter and Gummel (1969), an interpolation which ensures a constant current density along mesh lines allows coarse meshes to be used without loss of accuracy. Physically, provided generation is not large within an element, it is not unreasonable to expect the current density to be constant. This interpolation is used in all commonly available simulators.

5 THE SOLUTION OF THE EQUATIONS

This is not the place to enter into an extended discussion on the variety of techniques used to solve the large set of non-linear equations that arise from the discretisation of the device equations. However for the user to get an understanding of the limitations to the accuracy of the simulator a few concepts must be discussed.

Firstly, the equation set is highly non-linear. This arises because, under equilibrium conditions the carrier density depends exponentially on the potential through the Boltzmann statistics. Thus:

$$p \propto \exp(-q\Psi/kT)$$

This implies that even for changes of potential as small as 25mV (kT/q at 25°C) the equations can exhibit non-linearity.

Secondly, since the equations are non-linear, an iterative procedure is required to solve them. Thus an initial guess is needed for the solution. This guess does not satisfy the equations exactly, and so in light of the discrepancy the potential and carrier densities are updated. This procedure is repeated and the exact solution is approached at each update. This poses the question, when can we stop this iteration process and say we are happy with the solution we have? Clearly truncating this process too early will introduce an error whilst delaying unnecessarily costs computer time. An example of this trade off will conclude this section.

Two iteration procedures are commonly used in the device simulators, and are illustrated in figure 5. The fully coupled method, or Newton method, takes all the equations together ($\underline{F}(\underline{u})$) and uses the derivative of these equations with respect to the variables ($\underline{u}$) to guide the update procedure. Mathematically this can be obtained by truncating a Taylor series; thus it is required that:

$$\underline{F}(u + \delta\underline{u}) \approx \underline{F}(\underline{u}) + \partial\underline{F}/\partial\underline{u}\ \delta\underline{u} = 0$$

hence $$\delta\underline{u} = -\ [\partial\underline{F}/\partial\underline{u}]^{-1}\ \underline{F}(\underline{u}) \qquad 12$$

Note that as there are n equations to be solved (three for each mesh point) and n variables, $\partial\underline{F}/\partial\underline{u}$ is an n by n matrix. For typical problems of ≈1000 mesh points solving 12 is expensive. The fully coupled process requires a good initial guess from which to converge, but once the updates in potential are less than 25mV, rapid (quadratic) convergence is guaranteed and machine accuracy can be achieved in a further 4 iterations.

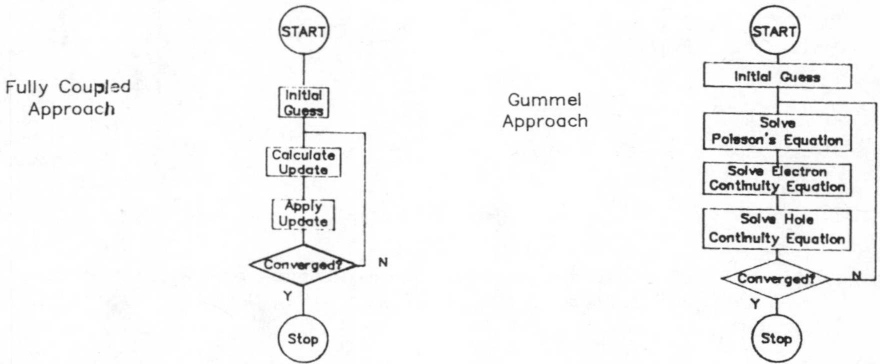

Figure 5. Iterative schemes employed by device simulators

The alternative procedure is due to Gummel (1964) and bears his name. In this procedure the equations are solved sequentially. However this ignores coupling between the equations, so the procedure is repeated until convergence. Because this procedure does not generate such a large matrix to invert, it is more economical per iteration than the Newton procedure. Moreover it converges better from a poor initial guess. However, it does not display quadratic convergence at the end of the iteration and is therefore more difficult to terminate.

Having outlined the different iteration techniques used, it is time to return to the question of truncation of the iteration sequence. It is possible to look at the magnitude of the function $\underline{F}$ at each iteration and stop the iteration when this becomes sufficiently close to zero. However this function does not relate simply to the accuracy criteria that the user might have, for example that the terminal currents are accurate to 1% or that the carrier densities are accurate to 10%. It is therefore normal to watch the updates in the carrier density or change in terminal current at each iteration and terminate the sequence when they become small enough. For the Gummel procedure this can lead to an unexpected loss of accuracy due to the slow convergence of the procedure. (Just because the present iteration has updates less than 1% does not mean that the next 10 iterations will not have similar updates leading to an error close to 10%!)

This is illustrated in figure 6 using a simulation of a MOSFET. The effect of demanding a smaller change in terminal current before truncation of the sequence, on simulation time and final terminal current is

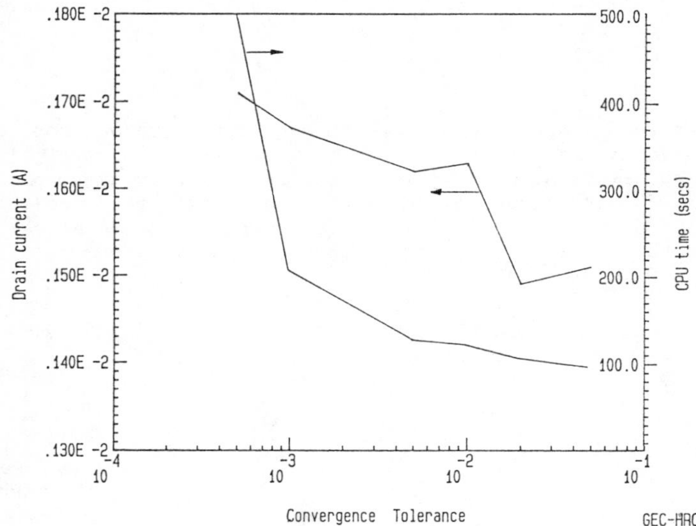

Figure 6. ffect of varying convergence tolerance on simulation time and 'converged' current for a simulator using Gummel's iterative scheme

presented. Clearly, as a tighter tolerance is demanded more iterations of the Gummel loop are required and the time increases. More surprisingly, as the tolerance is reduced from 5% to 0.5%, the terminal current changes by some 13%. This illustrates the danger of interpreting convergence tolerances as accuracies. If in doubt, the best course of action is to monitor the effect of convergence tolerance on the parameter of interest to you in your simulation.

6 CURRENTLY AVAILABLE DEVICE SIMULATORS

In this section a list of the commonly available device simulators is given. It does not include the many simulators that are reported in the literature but are only available 'in-house'. Where I am aware of the details of the operation of the simulator, its operation is summarised in Table 1.

7 THE MODEL FOR MOBILITY

To be able to accurately simulate semiconductor devices, not only must the numerical errors be kept under control but the physical models must also be of known accuracy and applicability. In this section a mobility model which can be applied both in the bulk of the silicon and at a (100) surface is presented. Whilst the bulk model is supported by data in the literature, the surface model is supported by a wide range of measurements on MOSFETs. Whether the mobility at a silicon surface is an intrinsic property of that surface, or whether it depends critically on the preparation of the surface, is still an open question; the data presented here suggests that the mobility does not vary excessively between devices manufactured by different processes. However the variation is sufficient to limit the absolute accuracy of device behaviour predicted by a simulator whose mobility model has not been calibrated to the technology.

TABLE 1.

	MINIMOS	BAMBI	PISCES	H FIELDS	BIPOLE
mesh	F. D.	F.Boxes	F. Elem	F. Elem	quasi 2D
Adaptive meshing	yes	yes	yes	no semi-automatic	?
solution scheme	Gummel	coupled	coupled	coupled	own approach
General device	no (MOS only)	yes	yes	yes	no (Bipolar only)
latest version	3.0	1.1	2B	1.?	?
features	widely used	very flexible	wide range of physical models prof. support interface with other products	range of contact & interface models	
time dependent simulations	no	yes	yes	yes	?
Availability	S. Selberherr Univ. Vienna		TMA inc.	Prof G Baccarani, Univ Bologna	D.Ralston, Univ. Waterloo

The model has been developed with the physics of the scattering mechanisms influencing the form of the model. Investigations of the strength of quantisation effects caused by the steep potential well which contains the carriers moving in a surface inversion layer indicate that, even at 300K, carriers do not behave as a 3d gas (whilst those carriers moving in the bulk of the silicon do). The dominant occupation of the lowest subbands is shown in Figure 7 for a typical n-channel FET. This will remain true whilst the drift field does not significantly heat up the carriers and allow occupation of the higher subbands. For this reason, the surface model presented here for mobility here has been developed independently of the bulk model. The transfer between the two models occurs smoothly as the surface is approached, using either the electric field or the distance from the interface as the controlling parameter. The form of the model is illustrated below:-

$$\mu(E_\perp, E_{||}, N_{dop}, T) = f_\perp(E_\perp).f_{sat}(E_{||}).f_{imp}(N_{dop}).f_{latt}(T).\mu_o$$

where the mobility is dependent on the following:

$E_\perp$ & $E_{||}$: the components of electric field normal to and parallel to the surface, respectively

N_{dop} : the total doping density

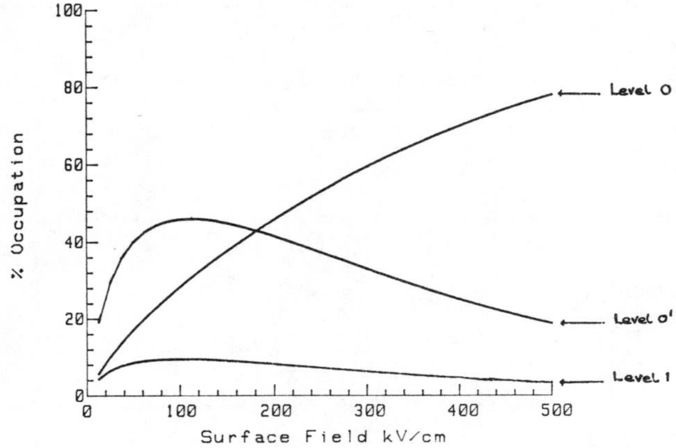

Figure 7. Occupation of sub-bands at a silicon surface showing predominant occupation of ground state at typical surface fields (at 300K)

T : the absolute temperature

Four mechanisms have been identified as responsible for the variation of the mobility from its base value of μ_o. These are:

a/ Lattice scattering

The carriers are scattered by the lattice vibrations (Phonons) whose density increases with temperature thus introducing a temperature dependence which is empirically found to be of the form:

$$f_{latt} = (T/T_o)^{-\lambda}$$

This term is included in both the bulk and surface mobility models. However the experimental measurement of the temperature dependence of the

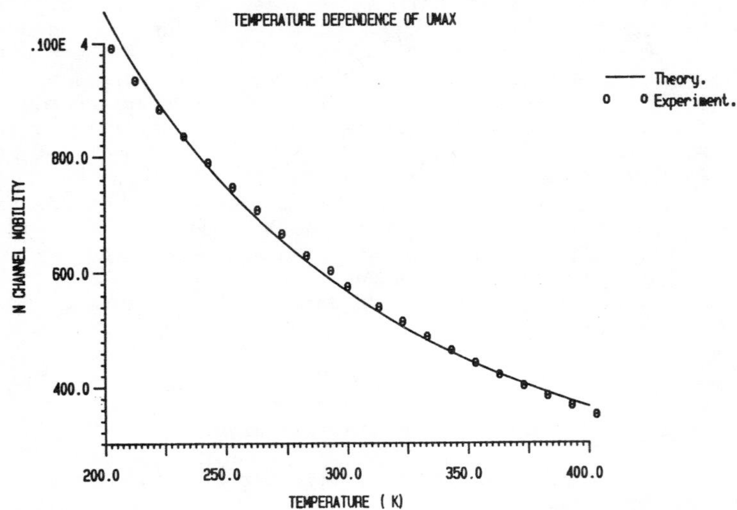

Figure 8. Temperature dependence of electron surface mobility

surface mobility (Figure 8) shows that the value of the temperature coefficient (λ) is about 1.5 compared to ≈2.5 in the bulk. This strengthens the case for separate treatment of the bulk and surface mobility models.

b/ Impurity scattering

The presence of a high density of charged impurity atoms scatters the mobile carriers through electrostatic interaction. Both positive and negatively charged impurities will scatter carriers so the total charged doping concentration (ionised donors plus acceptors) should be the controlling parameter in any mobility model. The form for scattering dependence in the bulk is due to Caughey and Thomas (1979) and is given by:

$$f_{imp} = A + (1 - A)/(1+(N_{dop}/N_{ref})^{\alpha})$$

$$\text{where } A = \mu_{min}/(\mu_o \cdot f_{latt})$$

Impurity scattering does not seem to be important in the channel of today's MOSFETs, although an increased impurity concentration will reduce the mobility through its effects on the surface electric field. However as channel dopings are increased above $10^{17}cm^{-3}$, a decrease in mobility is to be expected.

c/ Velocity saturation

When the carriers are moving in the presence of a high drift field, they acquire energy from the field more rapidly than they lose it by scattering; they therefore heat up. As they heat up more scattering mechanisms become available to them and as a result the velocity increases less rapidly with increasing field. Eventually the velocity saturates. This saturation is a function of the carrier temperature which is not usually available in today's simulators. However, the carrier temperature is determined by the drift field which is readily available and therefore the saturation is formulated in terms of this component of field (Canali et al, 1975), thus:

$$f_{sat}(E||) = 1/(1 + (\mu.E||/v_{sat})^{\beta}]^{1/\beta}$$

$$\text{where } \mu = f_{imp}(N_{dop}) \cdot f_{latt}(T) \cdot \mu_o$$

Careful measurement of the reduction of mobility as a function of drift field has been made for electrons moving at a 100 surface by a method developed from the ideas of Jerdonek and Bandy (1979). This shows that the choice of β as 2 in the above expression does indeed model the onset of saturation (Figure 9). Moreover, the method indicates that the saturation velocity is a weak function of both gate bias applied to our test device and temperature.

d/ Normal field dependence

Finally the mobility at the surface is found to be a function of the average confining field $E\perp$ (Sabnis and Clemens (1979)). This has been attributed to the reduction in the dimensionality of the carrier gas due

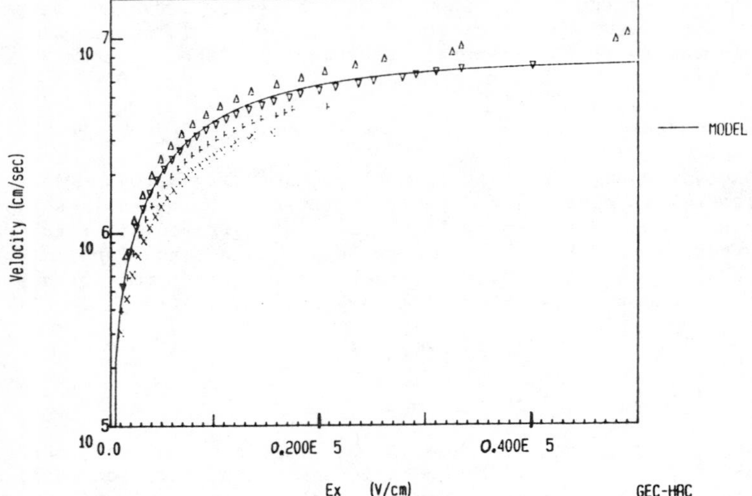

Figure 9. Electron surface velocity as a function of drift field for various gate biases on measured device (at 300K)

to the quantisation of the carriers in the surface potential well. The carriers are therefore scattered as if they were a 2d gas rather than a 3d gas when they are confined to a surface well and the scattering will then depend on the degree of confinement. This effect can be simply modelled by:

$$f_{\perp} = 1/(1 + \Theta\ (E_{\perp} - E_c))$$

The confining field is controlled by gate bias, back bias and surface doping; all three methods of control give rise to the same variation of mobility. Thus it is important that when a comparison of mobility between processes is made the effects of confining field are removed. Figure 10 compares the observed variation of electron mobility with confining field for devices made by two UK companies with those reported by Sabnis and Clemens (1979). Despite the technologies having gate oxides varying from

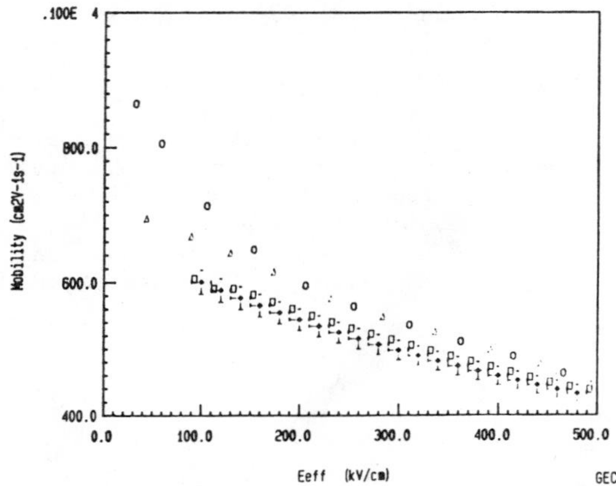

Figure 10. Variation of mobility with surface field showing similarity of mobility measured by Sabnis and Clemens [O,Δ] and at GEC

The parameters for the model discussed above are summarised in (Table 2).

Table 2. Table of mobility model parameters for (100) silicon surface

Parameter	Units	surface electrons	surface holes	bulk electrons	bulk holes
μ_o	$cm^2V^{-1}s^{-1}$	559	217	1448	473
T_o	K	300	300	300	300
λ	-	1.53	1.62	2.33	2.23
μmin	$cm^2V^{-1}s^{-1}$	not used	not used	92	47.7
N_{ref}	cm^{-3}	"	"	$1.3x10^{17}$	$6.3x10^{16}$
α	-	"	"	0.91	0.76
v_{sat}	cms^{-1}	$1.1x10^7$	$8.4x10^6$	$9.2x10^6$	$8.4x10^6$
β	-	2	1.21	1.92	1.21
Θ	cmV^{-1}	$1.0x10^{-6}$	$2.x10^{-6}$	n/a	n/a
E_c	Vcm^{-1}	$1.30x10^5$	$1.01x10^5$	"	"

30 to 100 nm the mobilities are in reasonable agreement. However the discrepancy does suggest that there may well be some process dependence of mobility which will prevent surface mobility from being treated as an intrinsic property of the surface.

Note from a numerical point of view that limiting β to integer values will improve calculation speed. We do not believe this will significantly affect accuracy; however it is still the subject of further investigation.

8 INTERFACING WITH OTHER SIMULATORS

The device simulator is only one of a family of tools that a modern VLSI process design engineer must use whilst developing a new process. Clearly, any device simulation package must rely heavily on process simulation for an accurate description of the device topology and doping. A process engineer will also want to know how well his transistors will perform in a circuit, thus a circuit simulator is another important tool. To date these tools have remained noticably separate. It is, however, interesting to ask the question; how closely do we require the device simulation software to interface with process and circuit simulators?

I can see no signs in the current trend, nor do I perceive a good reason for the whole process becoming a single integrated function where the engineer dials up a process recipe and gets a circuit speed out as a result of several hours of computation. However, obtaining a good transfer of data between process and device simulators is now becoming more important to all engineers who are growing tired of patching their favourite simulators together. Similarly, model extraction software and table driven simulators are becoming more common, allowing interface between device simulation output and a circuit simulator.

Currently, however, any new user of device simulation software should be warned that considerable effort is required to provide reliable interfacing between simulators and this should be borne in mind when first obtaining new software.

9 CONCLUSIONS

In this paper I have tried to present simply the sources of possible error when using a device simulator. I hope this makes users aware of the areas which need their careful attention. Firstly, many of the physical models underpinning the simulation have parameters which are extrinsic properties of silicon. If these are important to the operation of the device concerned they must be extracted for the material and process under development. Secondly, the physical models exclude investigation of the effects of propagating e.m. waves in the device or the effects of carrier heating (though there are currently many developments on the latter). Thirdly, the numerical techniques used commonly employ a finite mesh and iterative techniques, both of which introduce errors in the solution. It is really only possible for the user, bearing this in mind, to investigate the effect of these errors on his device simulation empirically. However, once this confidence in interpreting the results is obtained, the simulator offers a very flexible tool which can help the device engineer evaluate the complex trade-offs that occur when designing a modern device process.

REFERENCES

Adler M, Possin GT. (1981) IEEE Trans ED-28, 1053

Baccarani G., Wordeman M. (1985) Solid State Electronics 28 407

Blotekjaer K. (1970) IEEE Trans ED-17

Canali C. et al (1970) IEEE Trans ED-22 ,1045

Caughey D. and Thomas R. (1967) Proc IEEE 52, 2192

Chynoweth A. (1957) Phys Rev 109, 1537

Franz A. et al (1983) IEEE Trans ED-30, 1070

Gummel H. (1964) IEEE Trans ED-11, 455

Hall R.N. (1952) Phys Rev 87, 387

Hansch W. and Miura Mattausch M. (1986) J Appl. Phys. 60 , 650

Jerdonek and Bandy (1979) IEEE Trans. ED-25, 894

Kunhert R., Werner C., Schutz A. (1985) IEEE Trans ED-32, 1057

Sabris A., Clemens J. (1979) Confernece proceeding IEDM p18

Scharfetter D. and Gummel H. (1969) IEEE Trans ED-16, 64

Selberherr S. (1986) Simulation of semiconductor devices and process Pineridge Press, (vol 2).

Schockley W. and Read W. (1952) Phys Rev 87, 387

Sze S. (1981) Physics of Semiconductor Devices, Wiley-Interscience

Inst. Phys. Conf. Ser. No. 82
Paper presented at ESSDERC 1986, Cambridge 8–11 Sept. 1986

Monolithic microwave integrated circuits

F.A. Myers, Manager, GaAs IC Department

Plessey Research Caswell Limited, Caswell, Towcester, Northamptonshire.

1. Introduction

Modern microwave (3-30GHz) systems largely use a hybrid approach to the realisation of the system. This approach uses a discrete device, typically a GaAs FET, bonded into a package which is then mounted into a microstrip circuit to define the electrical function. This approach is directly analogous to the discrete transistor and printed circuit board technique common a few years ago for low frequency applications. At microwave frequencies the approach is capable of producing state of the art performance but can be labour intensive, larger than desirable and prone to failure unless great care is taken because of the number of operations carried out in the assembly.

The GaAs monolithic microwave integrated circuit (MMIC) potentially does for microwave applications what the Si IC has done for the lower frequency system. It puts all or most of the circuitry on a single chip of GaAs producing a small, lightweight, reliable and potentially low cost function block.

The paper will describe the present day state of MMICs and will deal with material, process technology, design, test and use of the chips in two complex microwave systems.

2. Material Growth

The basic building block of a GaAs MMIC is a GaAs FET. This device has been dealt with in detail many times and will only be briefly reviewed here. A schematic of the device is shown in Figure 1. The controlling action comes from the channel modulation effect of the Schottky barrier gate. The depletion edge successively pinches off the channel with increased reverse bias giving the potential for amplification at microwave frequencies. The frequency of operation of the FET depends upon the gate length. For example a 1 micron gate device can operate up to 10GHz, a 0.25 micron device up to about 40GHz etc.

For optimum operation, the quality and specification of the GaAs material is vital. It is necessary to fabricate a highly doped ($\sim 10^{17} cm^{-3}$) thin ($\sim 1 \mu m$) active layer onto a semi-insulating substrate, the boundary between the layer being as sharp as possible with a high active layer mobility. Four basic material technologies are, or have been, used to produce the active layer. These are vapour phase epitaxy, liquid phase epitaxy, ion implantation and molecular beam epitaxy. Chemical techniques, particularly vapour phase epitaxy, have been used extensively in the past, particularly for discrete devices and indeed are still capable of producing state of the art performance. However, for volume production, particularly of MMICs, the repeatability from layer to layer and variation across a layer in terms of doping and depth is not adequate. A major breakthrough has been accomplished in this respect by developing ion implantation into bulk grown substrate material as a viable approach for preparing the very thin active layers. Apart from some special applications, this approach is now the industry standard.

3. Fabrication Technology

A schematic diagram of a typical MMIC fabrication procedure is shown in Fig. 2. The critical steps that affect the MMIC performance and yield are the quality of the GaAs material, the gate and FET channel geometry, passivation, overlay capacitors, airbridges and vias (if used). For microwave circuits the active devices are usually operating in the linear region of their characteristics and, therefore, the detailed behaviour of the linear characteristic is critical. This is determined in a large part by the semiconductor doping profile beneath the gates of the FETs.

The gate process includes definition by photolithography down to 0.5 micron gate lengths and electron-beam lithography for shorter gates. The baseline industrial metal system for the Schottky gates is the Ti/refractory/Au metal system, used because of its proven reliability. The refractory barrier is typically TiW, Pt, W, Mo or TiW-Ta. In some processes, particularly for applications where 0.25 micron gate lengths are employed, low resistance gates are produced by Au plating the structures to form mushroom shaped cross-sections. Before the gate metal is deposited, recessing of the gate region is performed to achieve the proper value of saturation current.

The quality of the passive element processing can significantly affect performance and yield. Much attention has been given to:

(i) Deposition of a low-loss dielectric for metal-on-metal (MOM) capacitor formation;
(ii) Step coverage in non-planar structures;
(iii) High resolution patterning in the thick top metal which is performed by ion-beam milling;
(iv) Wafer thinning for controlled impedance transmission lines and
(v) Provision for via grounding through the substrate particularly for higher packing density and higher frequency circuits.

Dielectric deposition is restricted by the sensitivity of the ohmic and Schottky contacts to high temperatures following their formation. The most commonly used dielectrics are Si_3N_4, Ta_2O_5, SiO_2 and polyimide. Also important is the reduction of pinhole density in these layers particularly when the layers are thin. If the dielectric is used to separate first and second level metals, then the step coverage of the dielectric is important to avoid short-circuits at crossovers. For this reason polyimide is often used since it not only has a low dielectric constant (3.5) but is also deposited to a typical thickness of 1 micron in contrast to 0.2 micron for Si_3N_4.

For higher frequency circuits, two additional steps need to be controlled carefully - wafer thinning and etched via fabrication. The wafer needs to be thinned and metallised on the back to provide either a microstrip ground or a controlled parasitic image plane for lumped components. GaAs can be thinned down to 100 ± 5 micron by a combination of lapping and chemi-mechanical polishing although 200 micron is standard. The baseline technology for vias is a combination of wet chemistry and reactive ion etching to fabricate via holes through GaAs wafers. Reactive ion etching (RIE) is a much more directional process than wet etching and thus less sensitive to wafer thickness variations.

DC testing of chips, usually in the form of measuring parameters of standard test cells, is performed before thinning and scribing and separating into individual die. DC parameter values can give a good indication of the anticipated microwave performance but cannot provide detailed information. This point is discussed further below.

4. GaAs MMIC Circuit Design

Unlike silicon IC circuit design and simulation, GaAs ICs tend to have low packing densities, except at low frequencies, because of the need to avoid spurious coupling between both active and passive components. At frequencies below ~4GHz either direct-coupled techniques can be used, such as that shown in Fig. 3(a) or a combination of active-matching techniques with a limited number of passive elements, as shown in the example of Fig. 3(b). In both cases not only is chip size important but also the final yield of the circuit which is controlled by the number of components and their individual yields. Today, GaAs IC yields can range from only a few percent to greater than 50 percent. This is because the various stages of the process have finite yields in addition to the yields of the one or sub-one micron gates in the MESFETs. Yields and technology improvement programmes are currently aiming to improve both the performance and yield but it must be stressed that present technologies can be considered of production status.

At higher frequencies extensive use is made of transmission line matching as shown in the H-Band amplifier of Fig. 4. However, use is made of small tuning capacitors and resistors so that a full IC process is still needed. Chip area is restricted to a maximum of $25mm^2$ in order to obtain an acceptable yield per wafer. The typical area of the chips described in this article is $4\ mm^2$.

Prior to Calma layout and mask-making, extensive computer-aided designs of the chips are carried out by means of such software as SUPERCOMPACT(R) and Touchstone(R) together with a number of in-house programs. Much attention is now being given to the development of a sophisticated GaAs IC software package which will enable the engineer to produce complex ICs more efficiently.

In the past many GaAs ICs have been produced following the 'breadboarding' of the circuits using a conventional hybrid approach. Although this is a valid approach for certain circuits at the research stage, only simple concepts can be checked using such a technique and the use of the sophisticated programmes referred to above is essential. The programmes, however, sophisticated and complex are only as good as the data they are presented with to manipulate. Very extensive programmes of work have been carried out at Plessey and, presumably, at most other manufacturers to fully characterise the available components over as wide a frequency range and tolerance range as possible. Thousands of individual components (FET's, resistors, capacitors, inductors, transmission lines, etc.) have been measured, each over a wide frequency range and over many production batches. The quantity of the data is large and can only be processed using computers.

Modern CAD is now accurate enough that most linear circuits (e.g. amplifiers) can be expected to work on a single pass basis i.e. following CAD only one mask design and process batch need be carried out to obtain chips functioning to specification. Complex or novel circuits that stretch CAD to its limits may require several iterations.

5. Application and Examples of GaAs IC's

Since the early 1970's when only one or two organisations were involved in GaAs IC R & D the interest has now grown such that upwards of 80 organisations claim some design or manufacturing capability in GaAs circuits. IC processes genuinely have reached production status leading to merchant vending, foundry and custom design, as well as satisfying the in-house requirements that were often responsible for stimulating the venture.

An idea of the range of application areas for GaAs ICs is given in the list below. The potential here is vast and has led to enormous and variable market predictions. In the remainder of this article only the application of the 1-30GHz range (MMICs) will be considered.

APPLICATION AREAS FOR GaAs BASED ICS:

MONOLITHIC MICROWAVE (1-30GHz) ICS

- Satellite Applications
  - Navigational Satellites
  - Direct Broadcast TV (DBS)
  - On Board Systems
  - Communications
- Terrestrial Communications
  - Cellular Radio
  - Point to Point Links
- Radar-Land, Air and Naval
  - Conventional Radar
  - Phased Array
- Electronic Warfare
  - Broadband Surveillance and Jamming
- Instrumentation
- Missile Guidance

MONOLITHIC MILLIMETRE WAVE (30-300GHz)

- Smart Munitions
- Decoys
- Communications
- High Definition Radars
- Missile Guidance

HIGH SPEED DIGITAL

- Computing
- Instrumentation

INTEGRATED OPTOELECTRONIC CIRCUITS

- Communications

- Signal Processing/Control
- Data Conversion
- Communications
- Optical Signal Processing
- Advanced Microwave Generation and Control

5.1. Small Signal Amplifiers

The GaAs FET has found its largest area of application in the field of small signal, usually low noise, amplifiers. It is not surprising, therefore, that this has been the most intensively developed area for MMICs.

Fig. 5 shows an example of an amplifier at 1.5GHz for satellite navigation or DBS IF amplifier application. The chip employs 0.9 micron gate length devices and offers 18dB gain with sub-2dB noise figures, a specification for which it outperforms equivalent silicon amplifiers. Because of its relatively low frequency this feedback amplifier utilises lumped element matching i.e. inductance, capacitance and resistance can be easily identified as discrete elements. This amplifier should be compared with the LNA shown in Fig. 4 which operates at about 6GHz and utilises 0.5 micron FETs. Because of the higher frequency and hence shorter wavelength, this device uses distributed transmission lines to realise the desired impedance transformations. The performance of this particular chip is similar to the 1.5GHz device.

5.2. Broad Band Amplifiers

Although reasonable bandwidth can be achieved with conventional matching techniques, the travelling wave or distributed amplifier has shown the greatest performance. In this technique the FETs are distributed along artificial transmission lines, the otherwise troublesome device parasitics being absorbed into the values of the transmission line elements. An equivalent circuit of such an amplifier is shown in Fig. 6, with the MMIC realisation in Fig. 7. This chip produces 5dB gain from 2 to 12GHz with a noise figure of ~6dB. This is very useful performance for electronic warfare applications. By increasing the size of the FET, examples of this type of amplifier have produced 1W of CW output between 50MHz and 4GHz without having to resort to complex heat sinking technologies as required in genuine power FETs.

5.3. Mixers

Most signal processing is carried out by Si circuits at I.F. frequencies and it is then required to mix with a local oscillator (see below) to generate an I.F. Commonly, this is done with Schottky diodes but can readily be carried out using FET-based MMICs, the non-linearity of the FET performing the mixing function. Fig. 8 shows a single ended mixer at 3GHz offering 7dB of conversion gain, a very useful figure in a gain budget.

A more complex mixer is discussed in 8.2.

5.4. Oscillators

Generation of a stable frequency is usually required to complement the mixer and amplifier as a local oscillator, or as a signal generator. This is a very difficult non-linear design area but is now being tackled seriously using MMICs. Fig. 9 shows an example of a 2-4GHz voltage controlled oscillator (VCO). In this case negative resistance and tuning is accomplished by Schottky barrier varactors in the gate and source of the FET. Power of around 30mW at 20% conversion efficiency can be readily achieved over octave bandwidths.

The Q-factor and hence stability of these oscillators is, however, low. For fixed frequencies dielectric resonators can be used or if, stable, broad bandwidths are required, synthesiser techniques can be used.

Only a small cross section of available functions have been described above. Switches, phase shifters, multipliers etc. have all been realised as MMICs and some of these can be seen below giving the modern designer virtually all the building blocks he requires.

6. Testing of MMICs

DC testing, mentioned above, is used as a process monitor but can give a reasonable degree of confidence about a particular chip working or not. It cannot, however, give any information into VSWR, noise figure, gain, ripple etc. etc.

The ability to perform microwave assessments of GaAs MMICs while still on the wafer reduces significantly the subsequent effort and costs required with the conventional MMIC evaluation techniques. With confidence in the IC performance at this stage, advanced techniques such as the use of multi-component 'super-boards' may be adopted, thus increasing the yield and reducing the costs of integrated microwave sub-systems.

The system developed by Plessey over the last few years has been configured for maximum versatility such that DC characteristics as well as RF parameters e.g. S-parameters, noise figure, gain compression, etc. can be evaluated. To do this effectively it is necessary to impose some restrictions on IC layout e.g. standardisation of input/output pads Fig. 10 shows a schematic of typical RFOW automatic test equipment where equipment switching, apart from probe channel selection and RF route reversal, is performed in a standard rack separate from the wafer probe station. A desktop computer controls all the test equipment, sets the RF signal routing, drives the wafer probe step movements and manages the large quantity of data produced. To perform the on-wafer measurements, a grounded coplanar probe system has been devised to form a low loss, low VSWR transition between the coaxial measurement system and the approximately 100 micron coplanar IC features. The probe design ensures minimal common lead inductance and radiation during measurement. Prior to RF measurements the system is calibrated automatically using integrated coplanar calibration components on an alumina substrate having the same physical width as the ICs to be tested. The calibration is verified using other components both on the alumina substrate on the GaAs IC wafer to be tested.

After satisfactory probe alignment, calibration measurements are performed. Fig. 11 shows the DC and RF probes in position on a MMIC wafer during operation. The equipment is measuring an amplifier covering 2 to 4GHz used in the phased-array module described below.

If measuring noise figure, gain and all S-parameters across a band of nine frequencies, each IC requires 1.5 minutes including wafer movement. With over 270 measurable ICs on the wafer shown, approximately seven hours is required to complete the wafer test. These testing times show a massive saving in both time and costs in comparison with conventional techniques.

The vast quantity of data obtained can be displayed in any desired format; for the purposes of demonstrating the power of the technique the total chip yield [all parameters on a wafer map] is most illustrative. This is shown in Fig. 12 where, for the particular example, 57% of available die sites are electrically satisfactory.

7. Packaging

Packaging is not apparently at the forefront of technology and as such has not had the attention it deserves.

Bare GaAs chips can be handled by many organisations equipped with die and wire bonding equipment and in many applications this may be the ideal route. However, a bare chip is not very "user friendly" and some form of a convenient package is highly desirable. This must be cheap, easy to handle, robust and of good electrical performance. Three out of 4 of the requirements are easy to satisfy. Cost and electrical performance, i.e. maintaining impedance through a package wall with low insertion loss etc., tend not to be cheap. Few packages are available but more are being developed. A schematic of a typical example is shown in Fig. 13 which relies on 50 ohm stripline feeds through a ceramic wall to give the required performance.

8. Subsystem Applications

So far all the chips discussed have related to a specific function. In practice many microwave functions are required by a typical system and assemblies of chips have to be used. In this section examples of two complex subsystems will be described, one military, the other civil.

8.1. Phased-Array Radar Module

The concept of phased array radar is now well known. Simply, the radar generates its power by summing the outputs from many individual elements, the relative electrical phase between them controlling the beam width and position. In this way very high degrees of beam agility can be achieved with the ability to track and control multiple targets. However, to make a significant system impact, the individual elements have to be low cost in volume production (~£100's pounds each/10,000 off). GaAs ICs are starting to make a significant impact.

Fig. 14 shows the block diagram of the microwave function of an individual phased-array radar transmit/receive (Tx/Rx) module at ~3GHz. All the individual microwave functions except for the protection limiter are generated by GaAs devices, the three power stages being discrete GaAs FETs The phase shifters, Tx/Rx routing switch and LNAs are all GaAs MMICs.

The phase shifter, the heart of the unit, is an array of FET switches shown in Fig. 15. These switches control the lengths of transmission line and allow 2^4 phase states to be accessed. The LNA, shown in Fig. 16, is similar in function to, and half way in realisation between, the two LNAs described earlier. These devices, including the power FETs,

Tx/Rx switch and limiter are assembled into the complex microwave hybrid shown in Fig. 17. It should be noted that a complex Si control hybrid occupies over half the space. Each GaAs chip requires 2 Si chips to control it.

These complete Tx/Rx units have been built and tested and will make a large impact in radars, initially for military applications but hopefully for civil ones when prices start to fall.

8.2. Satellite Transponders

In order to accommodate the growing volume of communications traffic, future satellites will make increasing use of the allocated 6-4GHz frequency bands. As a result, these satellites will carry a large number of receivers, each assigned to geographically or polarisation isolated feeds. For this to be realised a new generation of miniature microwave receivers will be required. The emergence of GaAs monolithic microwave integrated circuits (MMIC) technology makes possible the development of receivers of small size and weight, low power consumption and with radiation-hard characteristics. The design, MMIC implementation and packaging of a 6-4GHz transponder currently under development, incorporating a low noise amplifier, mixer, miniature filter, high gain IF amplifier and integral power supply is briefly described.

The block diagram of the unit is shown in Fig. 18. Unlike the Tx/Rx unit this is a receive-only unit but with the requirement to down convert the nominal 6GHz signal to 4GHz, i.e. a mixer is required. This MMIC mixer is shown in Fig. 19. The input half of the circuit consists of two mirror imaged common-source FET mixers with input matching networks realised with high impedance transmission lines and overlay capacitors. The FET drains are combined in a single IF matching network to provide balanced operation. The mixer has a measured conversion loss of 3dB and a noise figure of 14dB. The noise figure is not critical as the device is preceeded by many stages of gain provided by several of the LNA chips shown in Fig. 4.

In order to correctly apply the local oscillator input (externally derived) an active splitter chip has been used as shown in Fig. 20. With a single input the device provides 10dB gain/channel balanced to 0.2dB. Additional phase shifting for optimum signal combining of the mixer is provided by passive MMIC chips using hi-pass/lo-pass networks and providing the necessary routing and combining functions.

Low cost, although important, is not as important for a space based application as reliability and performance. The assembly route chosen consists of a milled aluminium, gold plated box. The box is in two sections; the top section holds the RF modules and the bottom section holds the DC power regulator and distribution pcb. The lid contains RF absorbing inserts to suppress any box resonances.

The MMICs are mounted on four tungsten copper gold plated carriers with signal routing on alumina microstrip line. DC bias is supplied via vertically mounted feedthroughs from the distribution pcb. The various MMIC chips in the assembly are shown in Fig. 21.

The prototype unit shown in Fig. 22 shows a weight reduction of greater than 2:1 compared with typical hybrid designs and an approximately 4:1 reduction of volume.

9. The Future

It is first necessary for GaAs ICs to make a signficiant impact in present day systems, but future trends can be identified.

The first of these will be a higher level of integration on a single chip i.e. chips will become multifunction rather than single. A design example of a single chip phased array unit is shown in Fig. 23. This will lead to reduced assembly times, simpler packaging, improved reliability and reduced costs. This trend is likely to continue leading to mixed functions on a chip i.e. a combination of analogue, digital and perhaps optoelectronic functions on a chip. These aspects are likely to push material growth techniques further as heterojunction devices will be required for some of these functions.

Higher performance MMICs will also start making an impact. The most likely contender is an improved device in e.g. the High Electron Mobility Transistor (HEMT). HEMT offers superior performance to a MESFET for a given geometry and will probably be the device used for realistic millimetre wave applications, should they be required. This device will also, initially, require a large amount of effort in material and technology but offers considerable promise to the microwave engineer and will almost certainly make an impact. A direct comparison showing the advantage of a HEMT over a GaAs MESFET is shown in Fig. 24 and this is well worth working for.

10. Acknowledgements

This work has been supported in part by the UK MOD (DCVD) and the International Telecommunication Satellite organisation. The author wishes to acknowledge his many colleagues who have contributed data.

11. Reference

Many sources are available but the reader seeking more general background, and specific references, is recommended to read:-

"Microwave Field Effect Transistors - Theory, Design and Applications" by R.S. Pengelly, Second Edition (1986), Research Studies Press.

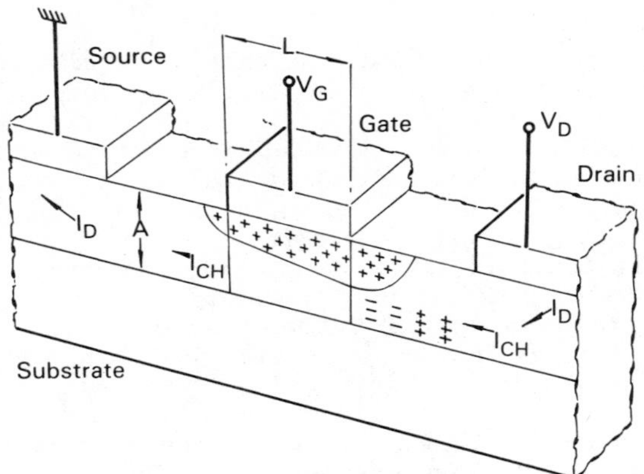

Fig. 1: Schematic of GaAs MESFET

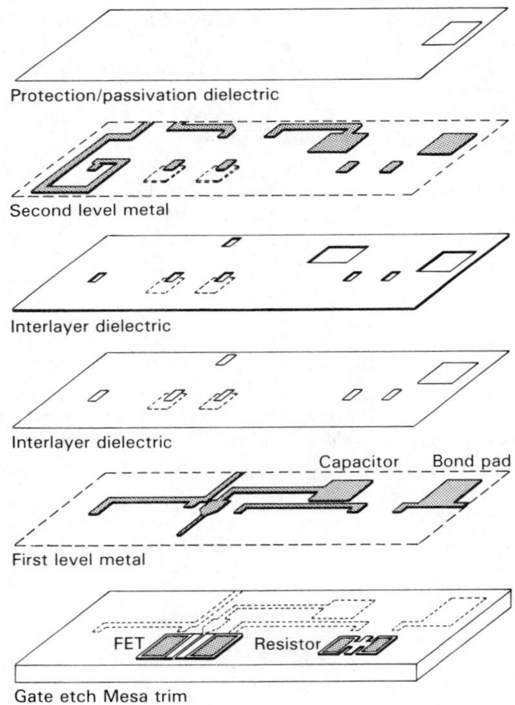

Fig. 2: Typical MMIC process

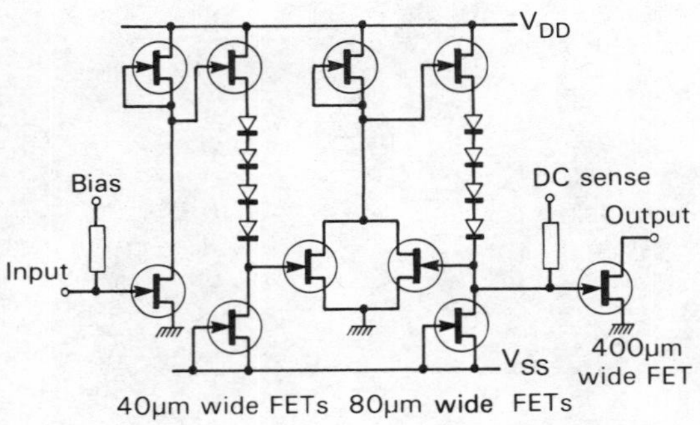

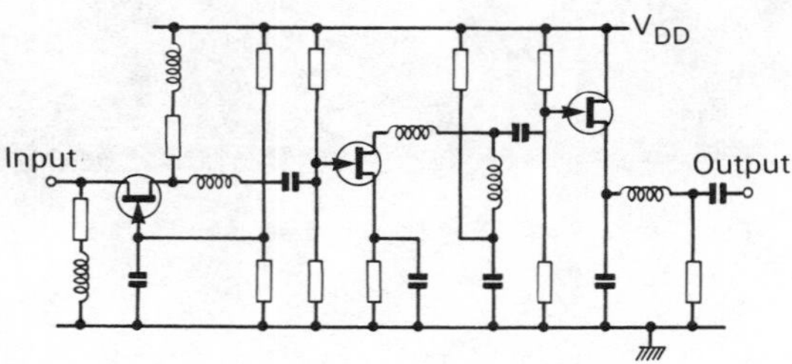

Fig. 3(a): Direct-coupled feedback amplifier

(b): Actively-matched broadband amplifier

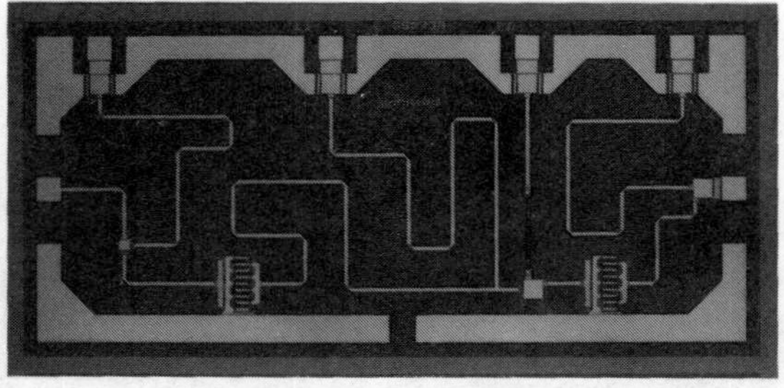

Fig. 4: Low noise amplifier (LNA) operating in the H-band (6 GHz) using transmission line matching

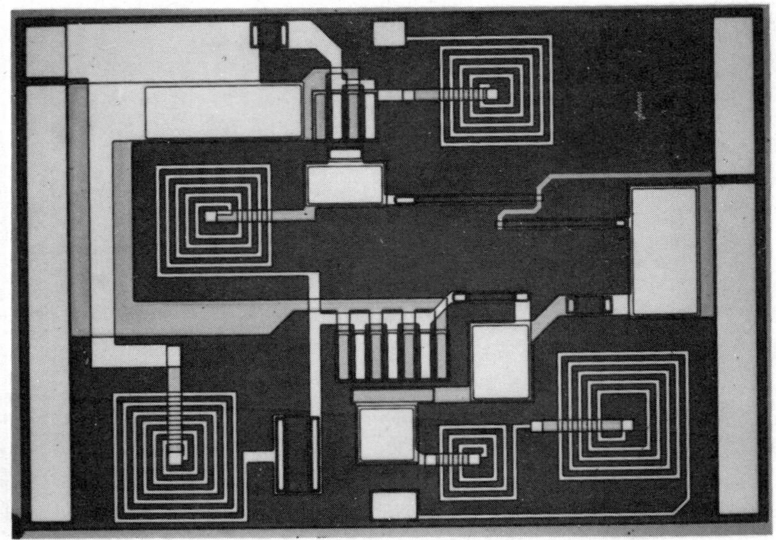

Fig. 5: 1.5 GHz low noise amplifier

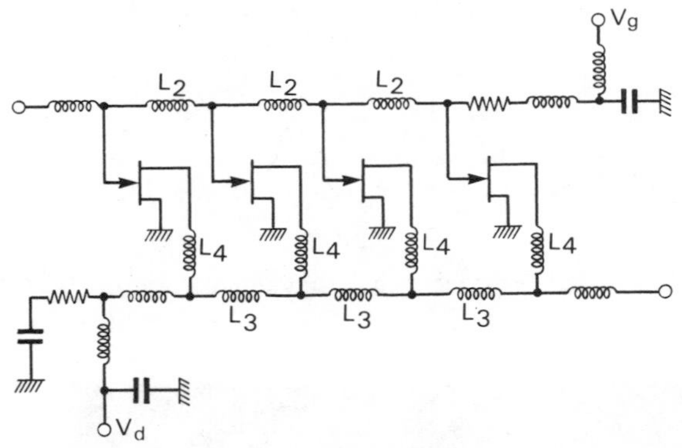

Fig. 6: 2-12 GHz travelling wave amplifier circuit

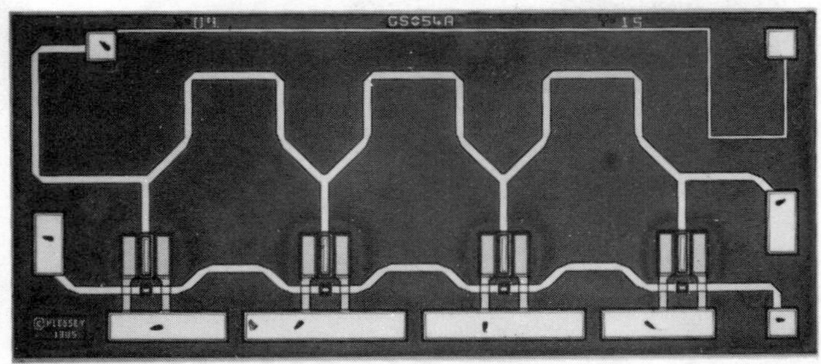

Fig. 7: Photograph of 2-12 GHz TWA chip

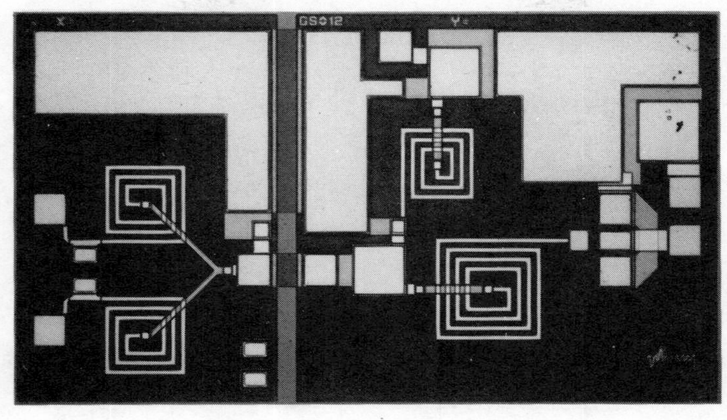

Fig. 8: Single-ended 3 GHz mixer

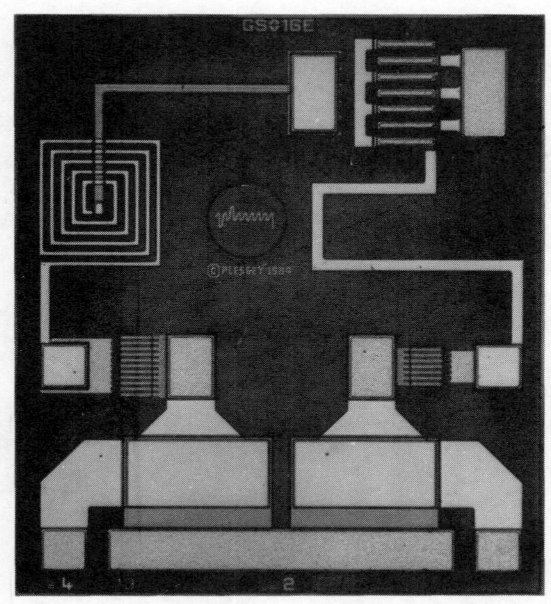

Fig. 9: 2-4 GHz VCO

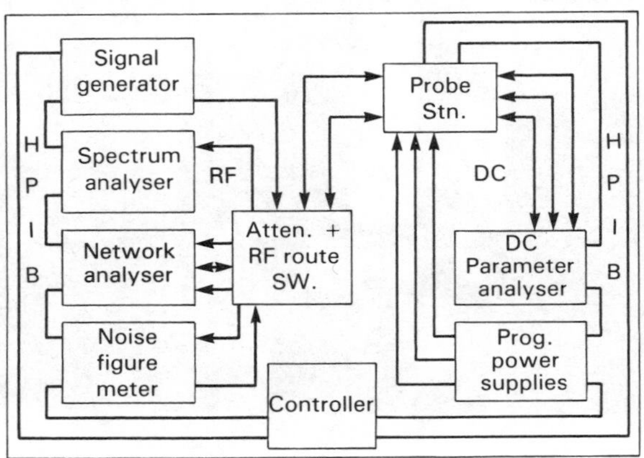

Fig. 10: Schematic diagram of RF on wafer measurement system

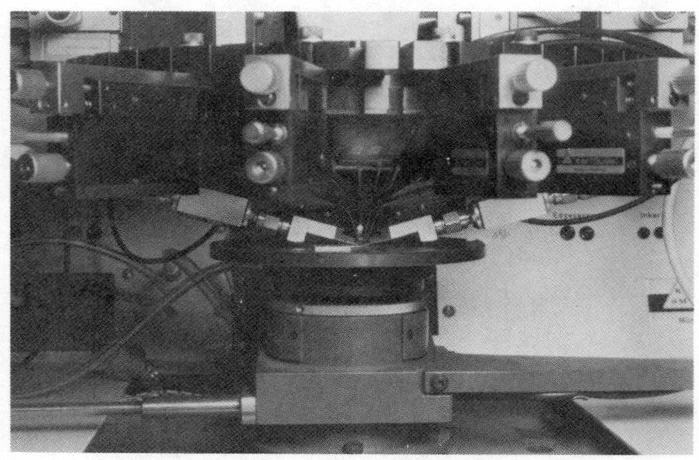

Fig. 11: View of probes measuring LNA chip

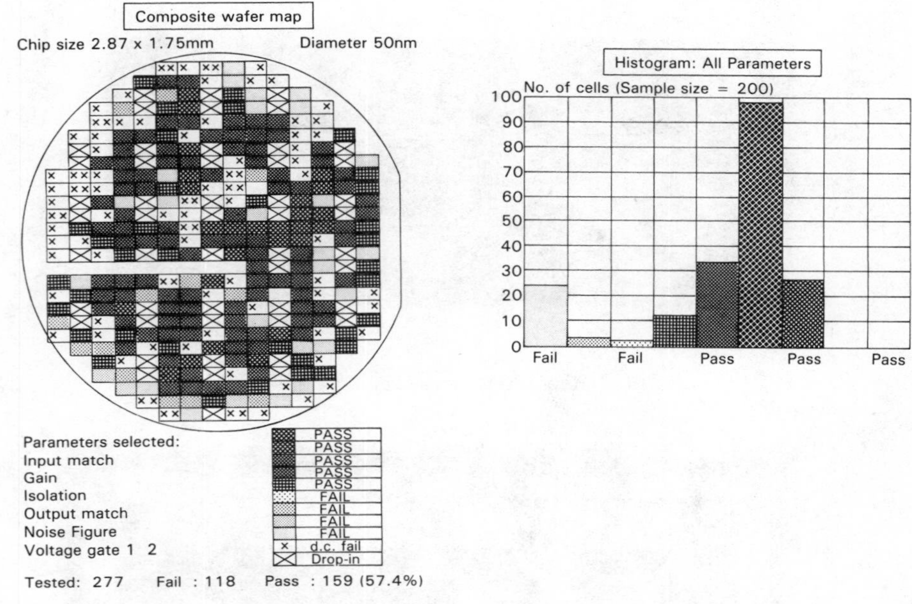

Fig. 12: Yield of LNA wafer

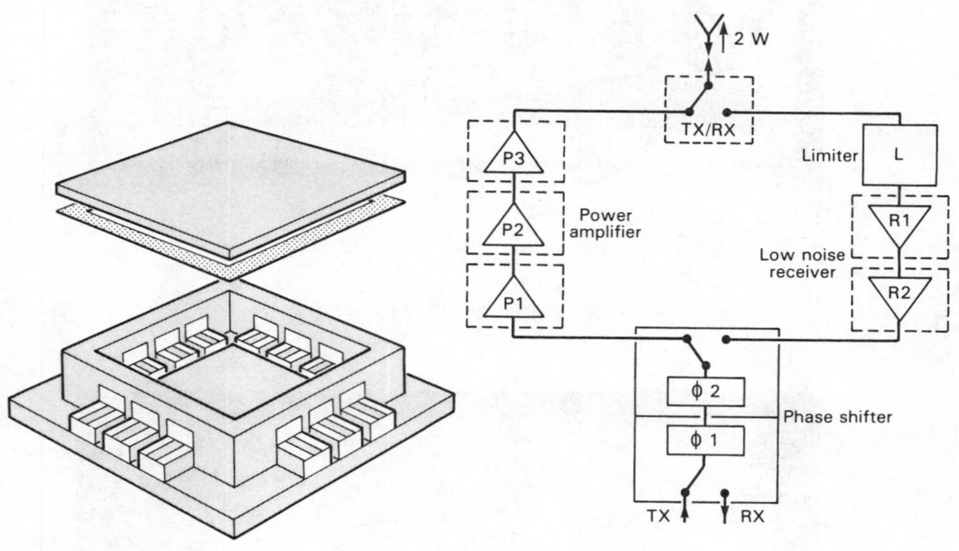

Fig. 13: Microwave ceramic package

Fig. 14: Block diagram of phased array module

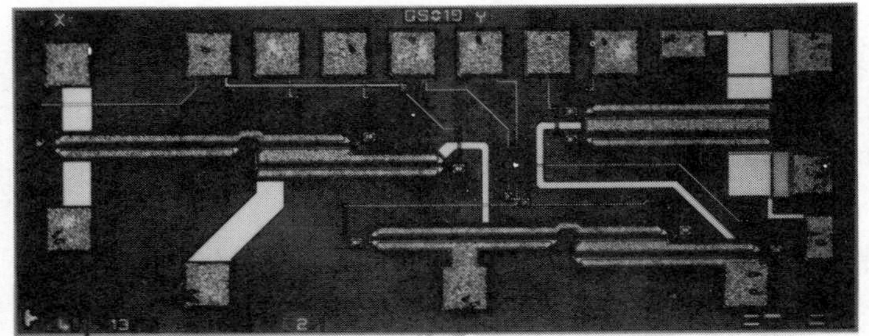

Fig. 15: Phase shifter switching network

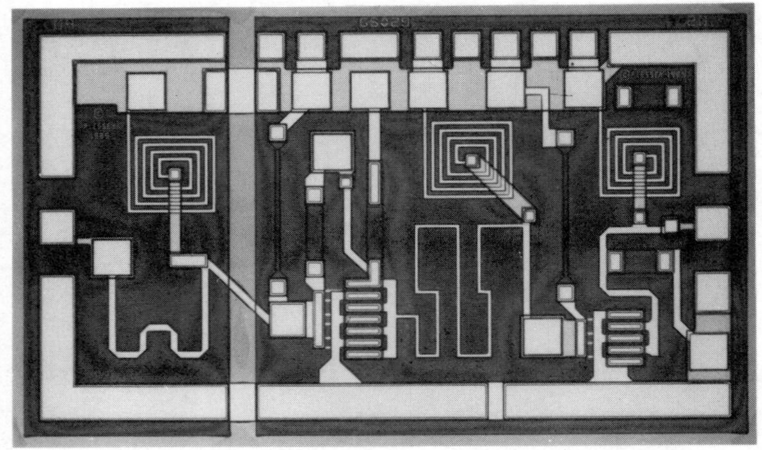

Fig. 16: Low noise amplifier chip

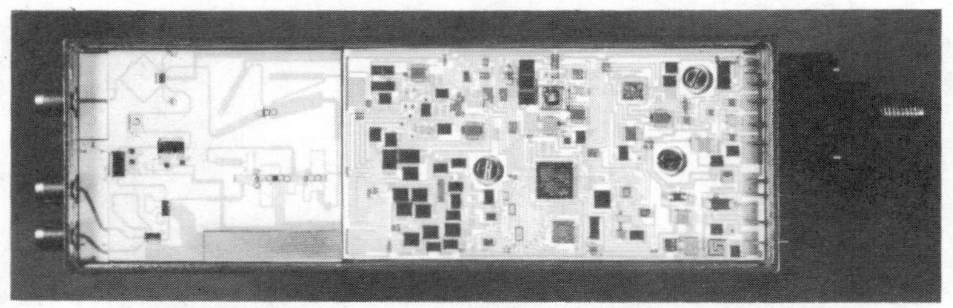

Fig. 17: View of overall phased array modules

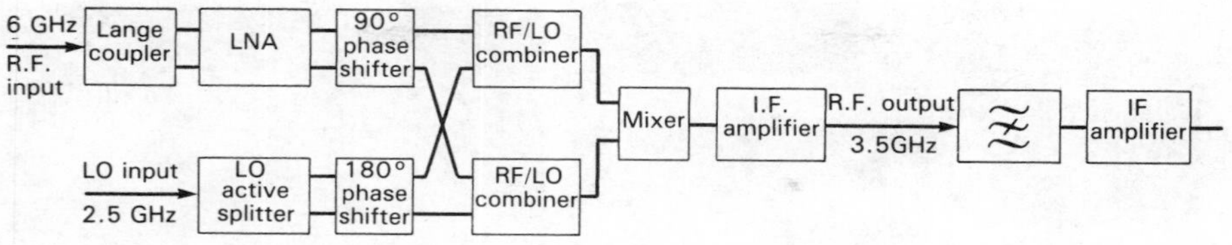

Fig. 18: Block diagram of 6/4 GHz transponder

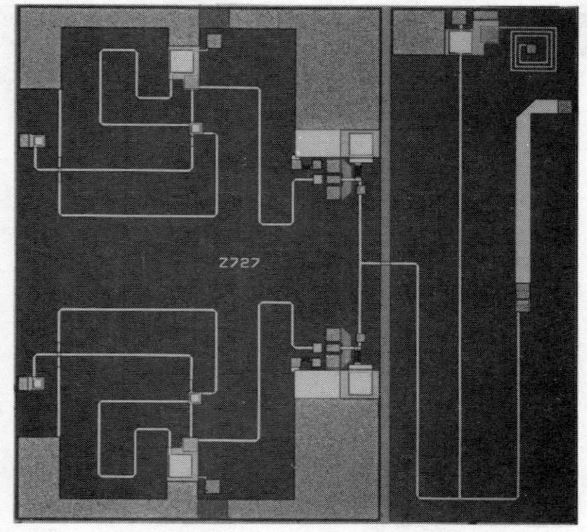

Fig. 19: Balanced mixer chip

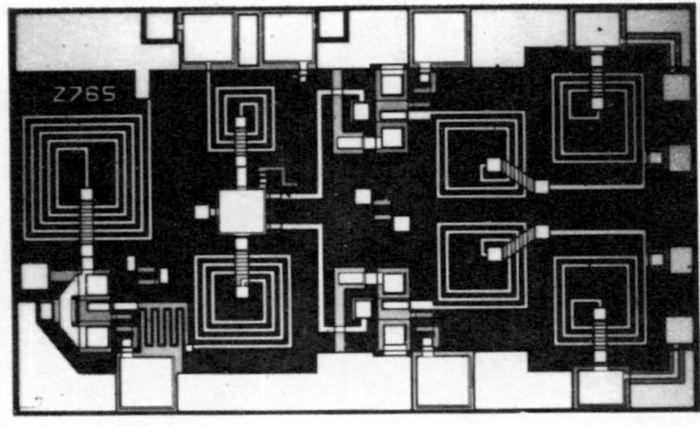

Fig. 20: Active splitter chip

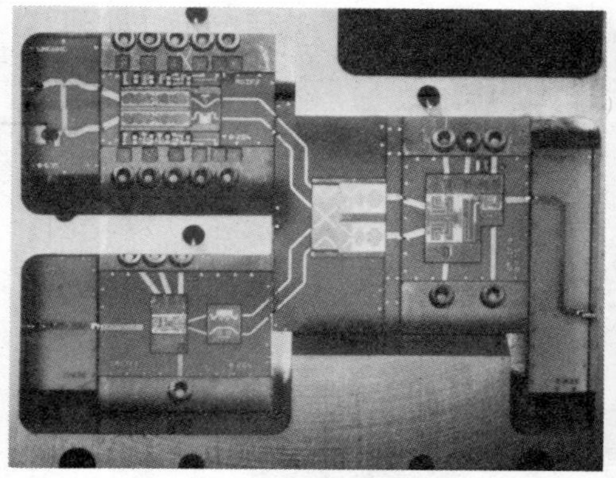

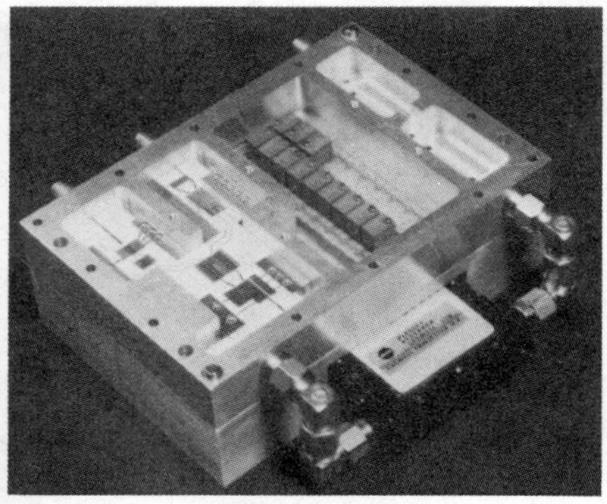

Fig. 21: Complete assembly of MMIC chips in transponder

Fig. 22: View of complete transponder

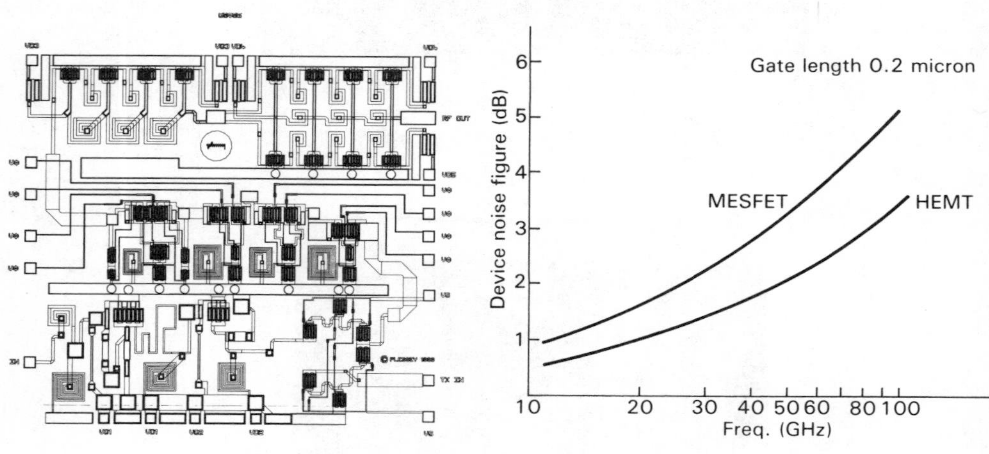

Fig. 23: Layout of proposed single chip module

Fig. 24: Comparison of GaAs MESFET and HEMT - noise figure/frequency

Inst. Phys. Conf. Ser. No. 82
Paper presented at ESSDERC 1986, Cambridge 8–11 Sept. 1986

Surface passivation for GaAs

Akihiro SHIBATOMI

FUJITSU LABORATORIES LIMITED

10-1 Morinosato-Wakamiya, Atsugi, 243-01, JAPAN

Abstract

In this paper, the development of passivation technologies for GaAs, such as native oxide and dielectric films, are reviewed. The applications of passivation films for GaAs device fabrication are described, including encapsulation for high-temperature annealing and inactivation films for GaAs devices. The effects of the overlayer stress on device performance to improve the short-channel effect, are also described.

1. Introduction

A very promising future is opening up for III-V compound semiconductors for applications in the fabrication of microwave analogue discrete devices, high-speed digital integrated devices and opto-electronic devices. A considerable amount of study has been done in the industry on developing the technology for the production of GaAs devices. By now, discrete GaAs metal-semiconductor field effect transistors (MESFET) as microwave high power and low noise devices, light emitting diode and laser diode as display or communication devices have a great success in the industrial fields. These devices are getting to be integrated for the purpose of monolithic microwave integrated circuits, memory and logic devices and opto-electronic integrated circuits. And also, new devices are emerging such as high electron mobility transistor (HEMT) and resonant tunnelling hot electron transistor (RHET) utilizing the properties and effects of hetero-junctions and quantum wells. These devices are supported by the recent advanced technologies of epitaxial growth method (Molecular beam epitaxy MBE, Metal-organic vapour phase epitaxy MO-VPE), ion implantation, dry fabrication processes etc, but the surface passivation technology for GaAs has been neither well established nor optimized. That is to say: the semiconductor-insulator interface technology. Without the establishment of this surface passivation technology of GaAs, we cannot expect to make high performance, high quality and reliable GaAs devices.

In this paper, the review of the recent development of the surface passivation technology for GaAs is discussed. In the paper applications of these technologies are limited to high-speed devices such as GaAs MESFET. In this device, the ion implantation technology is most important. First, encapsulation films over the GaAs at the high-temperature annealing step are described. The encapsulation over the GaAs MESFET devices such as discrete devices and integrated devices are discussed and the characterization of the interface states is described.

2. Review of surface passivation technology

Native oxide passivation is the key to the successful fabrication of modern silicon-based semiconductor devices such as MOSFETs. For GaAs devices, a suitable technology is necessary to form an insulator on GaAs with good dielectric and interface properties, particularly in view of the application to surface passivation and fabrication of various devices such as MOSFETs.

2.1 Oxidation

Similar techniques have long been sought to make native oxide on the surface of GaAs. However, there exists to date no dielectric films which would be obviously suitable for use with either of these elements. In order to provide a native oxide, while avoiding thermal degradation of the substrate, a number of techniques has been used to grow native oxide layers; thermal oxidation by Rubenstein /1/, Murarka /2/, and Butcher et al /3/, liquid chemical anodization by Hasegawa et al /4/ and Schwartz et al /5/, and plasma oxidation by Sugano et al /6/, Yokoyama et al /7/ and Chang et al /8/.

The growth method, composition and electrical properties of each oxide layer are summarized in Table 1.

Growth Method	Composition of Film	Resistivity (ohm-cm)	Breakdown Voltage (MV/cm)	Specific Inductive Capacity(εs)
Thermal-oxidation	$Ga_2O_3 > As_2O_3$	$10^9 \sim 10^{11}$	0.2~7	3~4
Anodic-oxidation	$Ga_2O_3 \simeq As_2O_3$	$10^{11} \sim 10^{16}$	3~7	5.4~8.7
Plasma-oxidation	$Ga_2O_3 \gtrsim As_2O_3$	$10^6 \sim 10^{15}$	0.5~6	5.4~8.5

Table I Growth method and properties of GaAs native oxide.

Thermal oxidation of GaAs in air was studied by Murarka /2/. The films were probably due to Ga_2O_3 and were amorphous, but these films were somewhat discouraging due perhaps to problems such as high insulator formation temperatures causing damage to the GaAs, for example, as vacancies which are related to the volatile components of As and As_2O_3. High surface state densities were formed in the interface. In order to improve the poor interface properties, anodic oxidation was carried out by Reves et al /9/ and improved anodic oxidation using a mixed solution of glycol and water (AGW) was carried out by Hasegawa et al /4/. This AGW process is an extremely stable, reliable, and reproducible process with which to form uniform, non-porous and glassy native oxide films with good adhesion. Plasma anodization of GaAs has been carried out by Sugano et al /6/, and Chesler et al /10/ in native glow region of a DC oxygen discharge, or by Yokoyama et al /7/ in the non-bias region of an RF excited oxygen plasma. The interface state density between p-type GaAs and its oxide film is of the order of 10^{10} $cm^{-2}eV^{-1}$ and between n-type GaAs and its oxide film is of the order of 10^{12} $cm^{-2}eV^{-1}$.

Although the discrete MOSFETs which used the oxide films were successful, these devices did not develop further in the industrial field because of the high density of the interface states and the unstable properties of the films.

2.2 Deposited films

Many kinds of dielectric films were developed as the passivation films for GaAs, which were deposited by many kinds of deposition methods such as chemical vapour deposition, sputtering, evaporation... etc. The deposition methods and properties of these dielectric films are summarized in Table II. In comparison with the properties of native oxide films, the deposited films have superior resistance in etching and stability in high - temperature processing, but have inferior uniformity, adhesion and density of interface states. All researchers report surface state densities in the 10^{12} $cm^{-2}eV^{-1}$ range along with electrical instabilities associated with the insulator and GaAs surface.

In spite of these instabilities, Beche et al /29/ were able to make an operating device using Si_3N_4 dielectric deposited films. But as with native oxide film GaAs transistors, the development of dielectric film GaAs transistors was not continued very long. Deposited films were only applied as encapsulation films for high-temperature processing, overlayer films for devices and the intermediate dielectric films between wiring levels. SiO_2 and Si_3N_4 films deposited by the CVD method or the sputtering method are typically used for the fabrication process of GaAs devices, as with Si devices. If the interface states of these films are

Deposition Film	Deposition Method	Resistivity (ohm-cm)	Breakdown Voltage (MV/cm)	Specific Inductive Capacity(ε_s)	Thermal Expansion Coeff. (°C)
SiO_2	CVD,Sputter	10^{6}	0.5~8	4.0~4.4	0.4×10^{-6}
Si_3N_4	CVD,Sputter	–	5~30	6~8	7.5×10^{-6}
AlN	Sputter,CVD	$10^{12}\sim10^{13}$	4~7	9.0	6.8×10^{-6}
SiOxNy	CVD,Sputter	–	5~10	–	6.8×10^{-6}
Al_2O_3	CVD	$10^{12}\sim10^{13}$	2~5	7.3~8	8.0×10^{-6}
Ge_3N_4	Evaporation	–	–	–	–
Ta_2O_3	Sputter	$10^{9}\sim10^{12}$	0.5~4	4~21	–
GaOxNy	CVD	–	–	5.5~6.0	–

Table II Deposition method and properties of dielectric films.

well controlled, these dielectric films are easily used in the fabrication process of GaAs devices because of the superior properties of the films chemically, physically and electrically. For the applications of these films to devices, the reduction of interface state densities is most important.

2.3 Theoretical treatment

An understanding of the origin of interface states is very important for the further development of GaAs-dielectric film interface technology and the possible removal of interface states. But, unfortunately, the origins of the interface states are not so well identified and the electrical properties of the interface not so clear as Si-insulator interface states. At the high-temperature annealing, thermal oxidation, or film deposition stages, chemical reactions exist between GaAs and atoms in the films, out-diffusion from GaAs, or/and in-diffusion from films, because of the high volatility of group-V atoms (As) and lack of stoichiometry. The point defects or stress-induced defects are introduced into the GaAs surface, which are believed to cause unwanted "pinning" of the surface Fermi level at the film-GaAs interface, and the interface states will appear.

Hasegawa et al /30/ summarized the electrical properties of GaAs metal/insulator/semiconductor (MIS) systems from various papers and discussed them from various points of view in an attempt to clarify the origin of the high density of interface states that characterizes the compound semiconductor-insulator interfaces. The explanations for the model of the insulator-GaAs interface state so far proposed can be classified into the following three categories, 1. The bonding defects in the transition layer between the semiconductor and the insulator give rise to interface states. Elemental arsenic at the interface and electron pairs in As_2O_3 have been proposed as the origin of interfacial traps by Chang et al /31/, Schwartz et al /32/, Breeze et al /33/ and Lucovsky et al /34/. 2. Interface states are defect-derived localized states which are introduced by the existence of the vacancies of Ga or As. The vacancies were established by the absorbed energy of metal or insulator atoms such as oxygen. This model is called the unified defect-derived state model by Spicer et al /35/. In fact, this is one of the remarkable advances in recent years in the understanding of compound semiconductor interfaces. 3. A unified disorder-induced gap state model for insulator-semiconductor and metal-semiconductor interfaces was proposed by Hasegawa et al /36/ /37/. Interface states are induced by defect-derived localized states at the interface.

Each model cannot yet explain completely the physics and chemistry of the insulator-semiconductor and metal-semiconductor interface. In spite of this fact, these interfaces are the basic constituent elements of modern GaAs devices.

3. Applications for GaAs devices

In recent years, much interest has been expressed in the use of GaAs for high-speed digital integrated circuits. Many ingenious device fabrication technologies have been proposed to realize ultra-high-speed GaAs logic circuits.

3.1 Encapsulation films for high-temperature annealing

In 1981, Yokoyama et al /38/ demonstrated the feasibility of using a self-aligned implantation technology using refractory gate metals for GaAs MESFETs to reduce significantly parasitic source series resistance so as to improve switching-speed performance. By using this technology, the large scale integration of logic and memory devices was developed by Nakayama et al /39/ and Yokoyama et al /40/, respectively. Now, refractory metal (TiW, WSi, WN, WAl...etc.) gate self-alignment technology has become a conventional fabrication process for GaAs ICs. Fig 1 indicates the four major stages in the fabrication of self-aligned GaAs MESFETs. First, a high-temperature-stable Schottky gate, which is formed by WSi_x deposition, is formed on an n-type GaAs layer. Second, a high-dosage Si^+ implantation is performed with the gate acting as an implantation mask. Third, dielectric films are deposited on the wafer and annealing is done at high temperature to activate dopants and to form the self-aligned n^+ region. Fabrication is completed by ohmic metallization. In this process, the depositions of the dielectric films are carried out after the ion implantations to form n-active layers and n^+-ohmic layers.

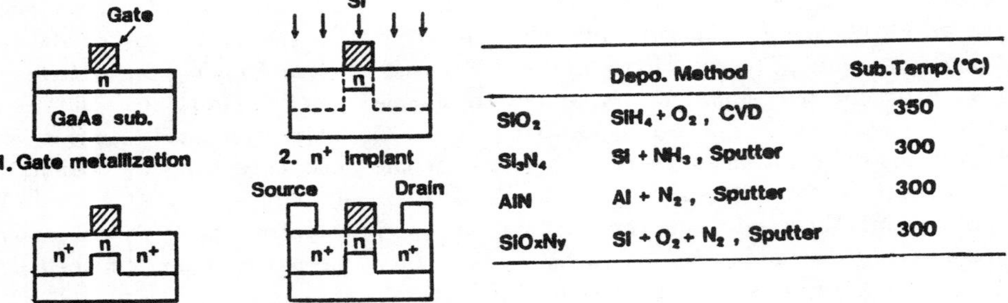

	Depo. Method	Sub.Temp.(°C)
SiO_2	$SiH_4 + O_2$, CVD	350
Si_3N_4	$Si + NH_3$, Sputter	300
AlN	$Al + N_2$, Sputter	300
SiO_xN_y	$Si + O_2 + N_2$, Sputter	300

Fig. 1 Major stages in the self-aligned GaAs MESFET fabrication process.

TableIII Deposition conditions of SiO_2, Si_3N_4, AlN and SiO_xN_y.

Reactive sputtered AlN is comparatively studied by Nishi et al /41/ with SiO_2 as an encapsulant for annealing of the implanted GaAs. The samples used for their study were Cr-doped semi-insulating GaAs substrates grown by the horizontal Bridgeman method. It is suggested that the interfacial strain arising from the thermal expansion mismatch between GaAs and the encapsulant plays an important role in redistribution of Cr and in determining the corresponding electrical properties of the implanted layer. And also they concluded that AlN is shown to be a promising encapsulant to minimize the effect of the interfacial stress which should be an important aspect for IC fabrication. But now, the semi-insulating substrates which were grown by the liquid encapsulated Czochralski (LEC) method are typically used for GaAs IC fabrications.

In the present paper, the optimization of the high-temperature encapsulant films on the undoped LEC grown semi-insulating substrate are comparatively studied with SiO_2, Si_3N_4, AlN and SiO_xN_y. The samples used for the present study were undoped semi-insulating 2" diameter GaAs substrates grown by the LEC method. The etch-pit densities were less than $2x10^4$ cm^{-3}. An implantation of 150 keV Si ions was carried out into a wafer to a dose of $2.2x10^{12}$ cm^{-2}. After the ion implantation, the wafer was cleaved into four pieces and each of these was encapsulated with SiO_2, Si_3N_4, AlN and SiO_xN_y of 100 nm thickness. The deposition conditions are shown in Table III. The resultant carrier profiles after annealing at 850°C for 15 min in the forming gas (5% H_2 + N_2), are shown in Fig. 2. A considerable deviation of the carrier profiles from the calculated one is observed in the case of SiO_xN_y annealing, in contrast with the results for AlN and Si_3N_4. After the annealing, each sample was measured by the Raman spectroscopy, shown in Fig. 3. The broadening of the L.O. phonon peak at 292 nm, observed in the sample annealed with SiO_2 and Si_3N_4 encapsulant, can be attributed to the interfacial strain arising in this structure as described by Nakamura et al /42/.

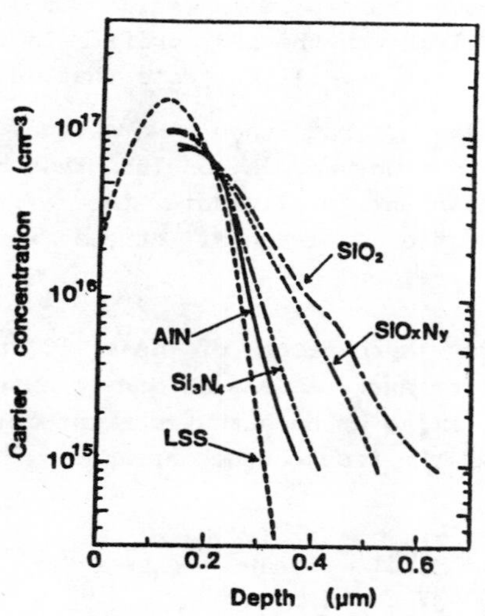

Fig. 2 Carrier concentration profiles for SiO_2, Si_3N_4, AlN and SiO_xN_y encapusulation annealing.

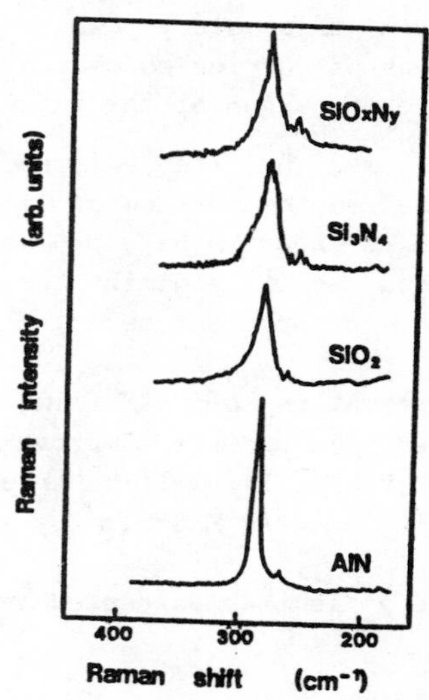

Fig. 3 Raman shift of SiO_2, Si_3N_4, AlN, and Si_xO_y encapsulated GaAs after annealing.

Deep tailing of the carrier profiles are observed in SiO_2 and SiO_xN_y capped samples. In another experiment, a more pronounced carrier tailing was observed after 200 min annealing at 850°C with SiO_2 and SiO_xN_y encapsulant, and a simultaneous reduction of the peak carrier density was observed. This result is thought to occur due to the gallium vacancy introduction from the SiO_2/GaAs and SiO_xN_y/GaAs interface as a result of Ga migration into the SiO_2 or SiO_xN_y encapsulant presented by Molnar

/44/, and Gyulai et al /43/. In the case of AlN and Si_3N_4 capped annealing, no such effects were observed even after annealing for 200 min at 850°C. Further experiments are necessary to confirm this effect.

In order to estimate the effect of stress between films and GaAs, ion implantations of 175 keV and dose of 4.0×10^{12} cm^{-2} were carried out to identify the peak values of each carrier concentration profile, and films of various thickness were deposited from 50 nm to 1000 nm. After annealing at 850°C, 30 min the peak values strongly depend on the thickness of SiO_2 and SiO_xN_y used for annealing, especially from 100 nm to 300 nm, but this dependence is weak for Si_3N_4 and is not observed for AlN. From this study, we can conclude that the main reason for the deviations of carrier concentration profiles from the LSS profile is the existence of oxygen at the interface, not stress. It suggests that Ga_2O_3 may be formed by the existence of oxygen, and then a Ga vacancy is introduced on the surface of GaAs. In this study we can conclude that the AlN film is shown to be a promising encapsulant to eliminate the effect of oxygen and to minimize the effect of the interfacial stress which should be an important aspect for IC fabrication.

Application of AlN films for the improvement of GaAs IC are described. In order to improve the performance of GaAs IC and to extend them to VLSI, it will be necessary to improve the output-current-drive capability of GaAs MESFETs by increasing the transconductance g_m.

The g_m is then expressed by,

$$g_m = \frac{\partial I_{dss}}{\partial V_{gs}} = c\epsilon\mu \frac{W_g}{dL_g} (V_{gs} - V_{th}).$$

From this equation, the common way to increase g_m is to reduce the gate length L_g, and to utilize materials with higher mobility μ. Another approach is to reduce channel layer thickness d, while maintaining the basic device dimensions. Onodera et al /45/ show that g_m can be significantly increased by employing an extremely thin channel layer, formed by the implantation of Si^+ ions into GaAs through AlN films.

The procedures for the fabrication of the FET's using the through-AlN implantation techniques are illustrated in Fig. 4. A 55 nm thick-AlN film was deposited by reactive sputtering onto semi-insulating GaAs substrates. Si^+ ions were then implanted at 59 keV through the AlN films, followed by annealing at 850°C for 15 min. During the annealing, the AlN film acts as an annealing encapsulant film. The process is continued to Fig. 4 (2). The doses of the through-implantation and of the conventional implantation were 1.2×10^{12} cm^{-2} and 3.6×10^{12} cm^{-2}, respectively, to achieve the same V_{th}. The experimental and theoretical carrier profiles were shown in Fig. 5. It was found that the through-implantated layer is very steep and also thinner than the conventional-implanted layer by approximately 50 nm, which roughly corresponds to the AlN thickness. The characteristics of both FET's were summarized in Table IV.

From our results, AlN films are promising encapsulants for GaAs.

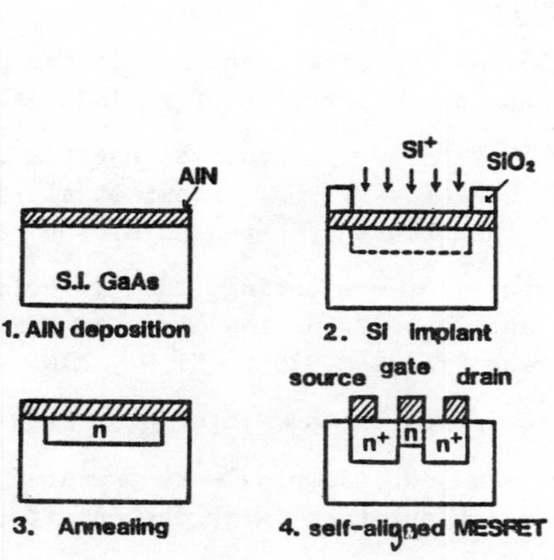

Fig. 4 Major fabrication steps of self-aligned GaAs MESFETs using implantation through an AlN layer.

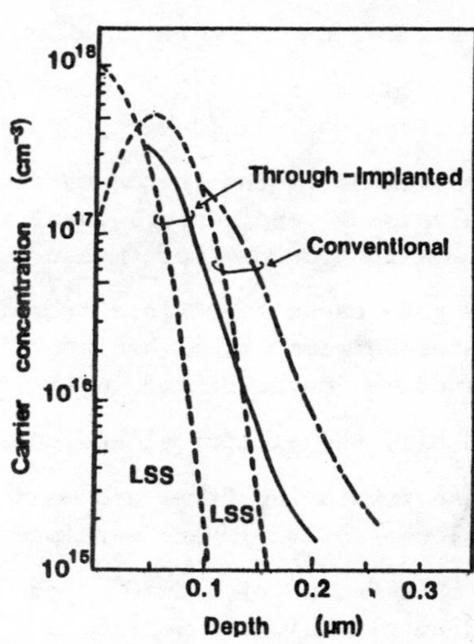

Fig. 5 Carrier profiles comparison for through-implantation and conventional implantaion.

	Effective layer thickness (nm)	Peak carrier conc.(cm^{-3})	g_m (mS/mm)	K-value (mA/V^2)	σV_{th} (mA)
Through-I/I FET	50	6×10^{17}	300	6.1	44
Conventional FET	100	3×10^{17}	190	2.9	65

L_g=1μm V_{th}=0.020V

Table IV Device properties of through-implanted- and conventional-implanted FETs.

3.2 Passivations for devices

The purpose of passivation in GaAs MESFETs is to improve the characteristics of the long term instability and leakage current of gate electrodes, which are caused by traps near the surface of GaAs. The formation of these traps is related to the chemicals and the surface treatments used during fabrication.

In this paper, we propose the optimum characterization method for interface states, and describe the interface properties of SiO_2, Si_3N_4, AlN and SiO_xN_y deposited films of GaAs MESFETs.

3.2.1 Application for discrete devices

Oldham et al /46/ showed that deep trapping centers in the p-n junctions of Si devices cause frequency dispersion of g_m, and that measurement of the frequency-dependence of this parameter is useful for analysis of deep centers localized in the active layer. Ozeki et al /47/ have also found that measurement of the frequency-dependence of g_m and the gate capacitance is extremely useful in evaluating the interface states between the insulator films and GaAs. Using the discrete MESFET structure encapsulated with insulating films (SiO_2, Si_3N_4, AlN and SiO_xN_y), the electrical and optical properties of the interface between these insulating films and GaAs were studied. Samples were prepared as follows; active layers were made by ion implantation with the conditions of 59 keV dose of 2.2×10^{12} cm^{-2} $850^{o}C$ 15 min annealing encapsulated with 100 nm AlN films. Source and drain ohmic metals were deposited and alloyed. Before the deposition of Al gate metals, the surface of active layers was etched off about a few nm to eliminate the surface damage layers.

Four kinds of dielectric films such as CVD-SiO_2, sputtered-Si_3N_4, reactive-sputtered-AlN and R-S SiO_xN_y were deposited on the samples with about 450 nm thickness. Measurements of frequency-dispersion of g_m were carried out for each sample. Fig. 6 shows the frequency dispersions of g_m for each sample. It was found that there are two types of frequency dispersion of g_m: one is highly dependent on the properties of the interface between the passivation film and GaAs, while the other is dependent on the electrical properties of the GaAs bulk crystal which are formed by ion implantation. The frequency dispersions A,B,C and D in Fig. 6 can be classified as surface-dependent because their magnitudes vary greatly according to the surface treatment and type of passivation films. The deviations using AlN films are very small over the frequency

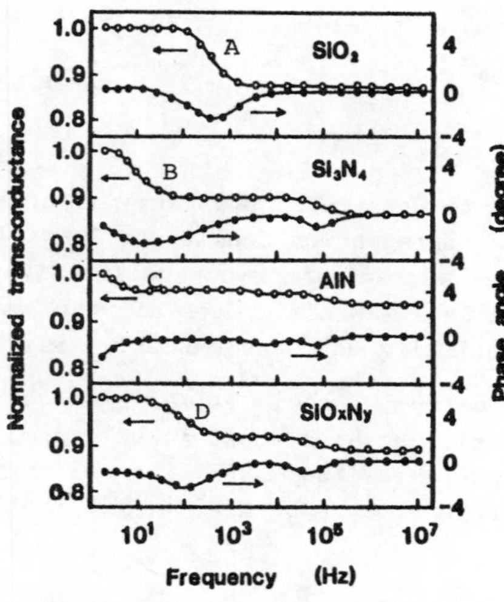

Fig. 6 Frequency dependence of transconductance in MESFETs.

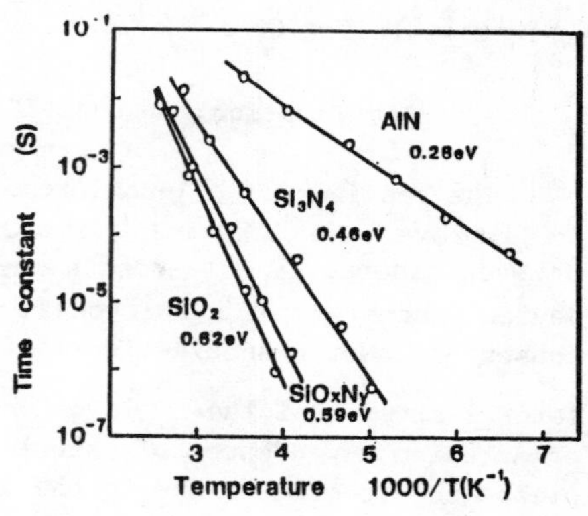

Fig. 7 Temperatur⌟ dependence of time constant.

range from few Hz to 10^7 Hz. The origins of these deviations were identified in detail by Ozeki et al /47/, as leakage currents at the interface, which were closely related to the frequency dispersion of both g_m and the gate capacitance.

The electrical and optical properties of the interfaces can be obtained from measurements of g_m and the gate capacitance. Fig. 7 shows the temperature dependence of the time constant for electron emission from interface states observed in four kinds of insulating films. It can be seen from this figure that each insulating film has a characteristic activation energy; the observed activation energy for SiO_2 film is 0.62 eV, that for SiO_xN_y is 0.59 eV, that for Si_3N_4 is 0.46 eV and that for AlN is 0.28 eV. The optical properties of the interface states were also investigated using the same sample. In the lower energy region, each passivation film has a characteristic spectrum; SiO_2 has a peak energy of 0.78 eV, Si_3N_4 has a peak energy of 0.99 eV and SiO_xN_y has a peak of 0.81 eV and AlN has a peak of 1.18 eV. The differences between these peak energies and the band gap energy are nearly equal to the thermal activation energy.

Device characteristics of each deposited film were also measured. The long term instability of the drain current and the leakage current of the gate electrode were also observed except for the AlN encapsulated devices.

AlN is shown to be a promising overlayer film to make a stable surface state for GaAs MESFETs.

3.2.2 Application for GaAs ICs

The realization of submicrometer-gate GaAs MESFETs is very important to improve the switching speed and integration density of GaAs IC. However, short-channel effects have been degrading the controllability of device parameters of submicrometer-gate GaAs MESFETs. One of the main causes of short-channel effects in self-aligned GaAs MESFETs is the lateral stretch of the n^+ layer observed by Ohnishi et al /49/. The orientation dependence of short-channel has been shown to be caused by piezoelectric effects due to the deformation of the MESFET channel region produced by a stressed dielectric overlayer by Asbeck et al /50/, Chang et al /51/, and Ohnishi et al /52/.

Ohnishi et al /52/ and Onodera et al /53/, my co-workers, proposed the possibility of utilizing the piezoelectricity of GaAs substrates to improve the performance of FETs. We characterize the influence of dielectric-overlayer thickness on threshold-voltage V_{th} and K-value of both [011] - and [01$\bar{1}$] - oriented WSi_x - gate self-aligned GaAs MESFETs in the cases of SiO_2 - and Si_3N_4 - dielectric overlayers. [011] - and [01$\bar{1}$] - oriented FETs, having five different gate lengths, were fabricated on a (100) 2" semi-insulating substrate grown by undoped LEC. The gate width was 20 µm. Si-implantation of the n-active layer was carried out at 59 keV with a dose of 1.0×10^{12} cm^{-2} and annealed at 850°C for 15 min. Self-aligned n^+-ohmic layers were ion implanted with 400 nm thick WSi_x gates used as masks at 175 keV with a dose of 2.1×10^{13} cm^{-2}, and annealed at 750°C for 15 min. Encapsulation films for both annealing processes were 100 nm AlN. Ohmic source/drain electrodes were fabricated as shown in Fig. 8. After the AlN films were etched off and the surface cleaned with some rinse, the wafers were cleaned into four pieces and 30 nm, 60 nm, and 120 nm CVD-SiO_2 and plasma-CVD-Si_3N_4 were deposited as dielectric overlayers.

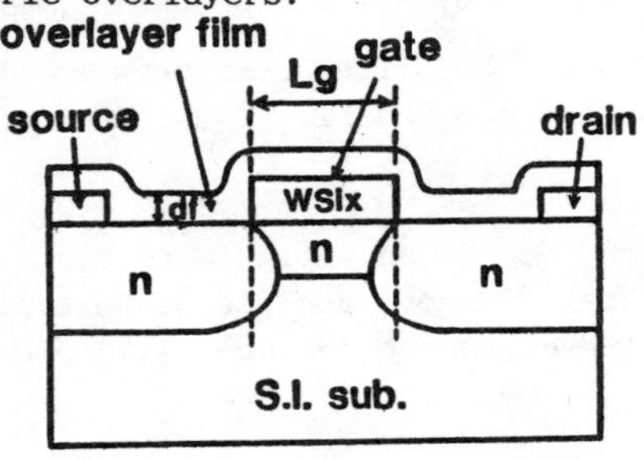

Fig. 8 Schematic cross section of the WSi_x-gate self-aligned GaAs.

The gate-length dependence of V_{th} of [011] - and [01$\bar{1}$] - oriented FETs, with SiO_2 - and Si_3N_4 - overlayer thickness d_f as a parameter shown in Fig. 9 and Fig. 10, respectively. For [01$\bar{1}$] - oriented FETs without a SiO_2 overlayer, V_{th} remarkably shifts to the negative side as the gate length is reduced to less than 1 µm, the same as for [011] - oriented FETs without a Si_3N_4 overlayer. On the contrary, no V_{th} shifts are observed as the length is reduced to less than 1 µm both for [01$\bar{1}$] - oriented FETs with a 120nm thickness SiO_2 overlayer, and [011} - oriented

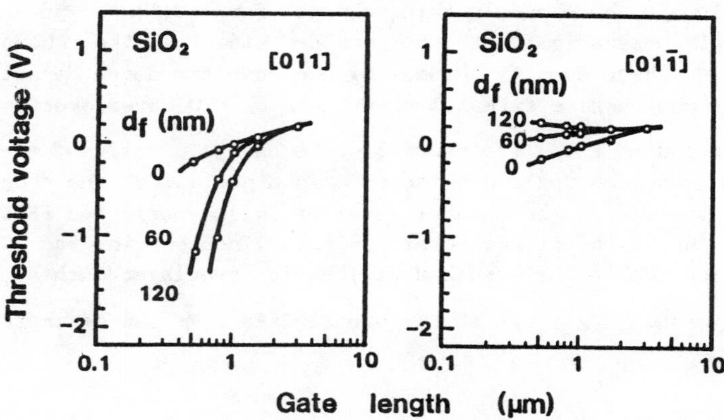

Fig. 9 Gate-length dependence of threshold voltage with SiO_2-film thickness, as a parameter.

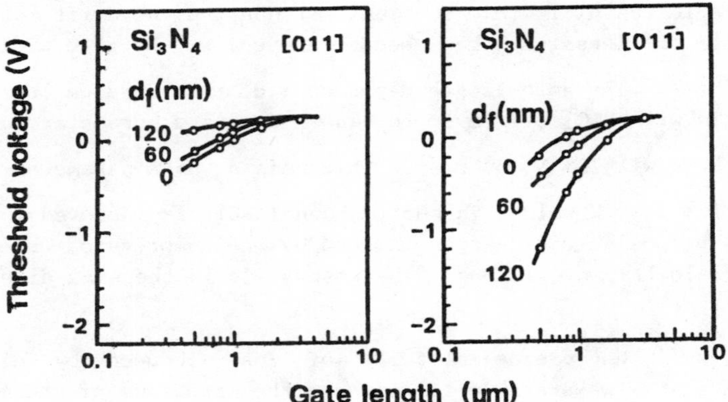

Fig. 10 Gate-length dependence of threshold voltage with Si_3N_4-film thickness , as a parameter.

FETs with 120nm thickness Si_3N_4 overlayer. These results can be explained by the modification of the effective channel thickness, due to the piezoelectric charge theoretically and experimentally by Onodera et al /53/, as follows; increasing the dielectric overlayer thickness, the stress under the gate region of the channel layer increases compressively or expansively, and piezoelectric charges are induced under the gate region.

So the piezoelectric charge density at the channel layer is proportional to dielectric overlayer thickness. From this the overall space-charge-distribution in channel layer is obtained from the donor atom profile and the piezoelectric charge for [011] - and [01$\bar{1}$] - oriented FETs. Especially at the deeper side of the channel, the piezoelectric charge density becomes higher than the donor density. For instance, if compressive stress such as a thick SiO_2 overlayer is applied to [011] - oriented FETs, the overall space-charge profile of the channel layer becomes more abrupt and thinner. In other words, the deep tail of the donor profile in the channel layer of [011] - oriented FETs can be compensated for by the piezoelectric charges induced in the GaAs FET channel region due to the stressed dielectric overlayer such as SiO_2 and Si_3N_4. As a result, g_m of short-channel GaAs FETs can be improved with smaller shifts in V_{th}.

The K-value is one of the criteria for FET performance and is expressed as

$$K = \frac{\epsilon \mu W_g}{2d L_g}$$

where ϵ is the dielectric constant of GaAs, μ the drift mobility, d the effective thickness of the channel layer, W_g the gate width and L_g the gate length. The gate-length dependence of the K-value of [01$\bar{1}$] - oriented FETs with SiO_2 overlayer thickness d_f as a parameter, and [011] - oriented FETs with Si_3N_4 overlayer thickness d_f as a parameter is shown in Fig. 11. The K-value in short-channel FETs is improved with the presence of piezoelectric charges induced by the compression or expansion stress of dielectric overlayers. The result is in the same direction as the V_{th} shifts.

By using the piezoelectricity of GaAs induced by dielectric passivation films, we are able to overcome the problems of short-channel effects in a self-aligned GaAs MESFETs.

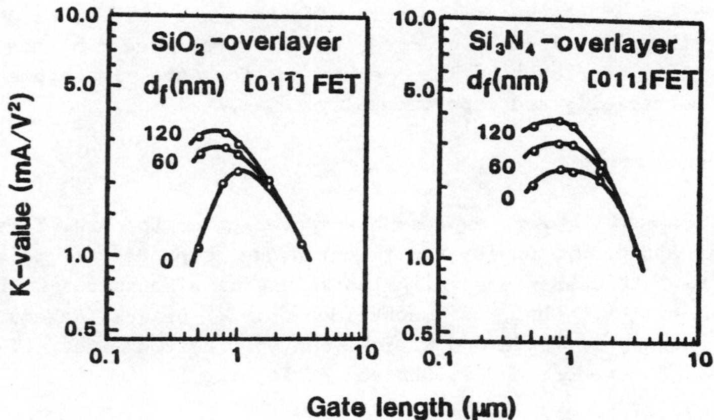

Fig. 11 Gate-voltage dependence of K-value with SiO_2- and Si_3N_4-film thickness, as a parameter.

4. Summary

In this study, the development of surface passivation films for GaAs was reviewed. Many kinds of GaAs native oxide films were tried but these did not develop so well, because of the existence of a high density of interface states and unstable properties. So the studies turned to the development of deposited films for surface passivation. This paper reviews the deposition method and properties of these films. But the properties of the interface between dielectric films and GaAs have not yet been well understood theoretically and experimentally.

The applications of these films as the surface passivation for GaAs devices were also described, which include encapsulation films for high-temperature annealing and inactivation films of GaAs surface and overlayer which induced the piezoelectric effect.

The properties of SiO_2, Si_3N_4, AlN, and SiO_xN_y films were compared and optimized to form high quality layers for high-temperature annealing as encapsulation films. AlN was demonstrated to be a most promising encapsulant for the future development of the GaAs IC, as the effect of the interfacial stress can be minimized, and the absence of oxygen reduces Ga out-diffusion.

For films to deactivate the GaAs surface, the same kinds of films were deposited, and these interface properties were characterized by the frequency-dispersion of g_m. The AlN film is promising for that purpose for the same reasons.

The problems of the short-channel effects were able to be overcome by utilizing the piezoelectric charges which were induced by the stress of the dielectric overlayer. The mechanism for the improvement was explained theoretically and experimentally.

5. Acknowledgement

The author would like to express his appreciation to Drs. T. Misugi, M. Kobayashi, and O. Ryuzan for their continuous support of this work. And also, I wish to acknowledge with thanks useful discussions with Profs H. Hasegawa and H. Ohno, of Hokkaido Univ., Drs. T. Nakamura, M. Takikawa, M. Ozeki, S. Inada and N. Yokoyama of Fujitsu Labs LTD. and Drs. H. Nishi, T. Onodera and T. Ohnishi of Fujitsu LTD.

6. References

/1/ M. Rubenstein; J. Electrochem. Soc. 113, (1966),540
/2/ S.P. Murarka; Appl. Phys. Lett, 26, (1975), 180
/3/ D.N.Butcher and B.J. Sealy; Electron. Lett. 13, (1977), 558
/4/ H. Hasegawa and H.L. Hartnagel; J. Electrochem. Soc. 123, (1976),713
/5/ B. Schwartz, F. Ermanis and M.H. Brastad; J. Electrochem. Soc. 123, (1976), 1089
/6/ T. Sugano and Y. Mori; J. Electrochem. Soc. 121, (1974), 113
/7/ N. Yokoyama, T. Mimura, K. Odani, and M. Fukuta; Appl. Phys. Lett. 2 (1978), 58
/8/ C.C. Chang, R.P.H. Chang and S.P. Murarka; J. Electrochem. Soc. 125, (1978) 481
/9/ A.G. Reves and K.H. Zaininger; J. Am. Ceram. Soc. 16, (1963), 606
/10/ L.A. Chesler and G.Y. Robinson; Appl. Phys. Lett. 32(1), Jan. (1978) 50
/11/ W. Kern and J.P. White; RCA Rev. 31, (1970) 771
/12/ L.G. Miners, P.Pu-Pin and J.R. Sites; J. Vac. Soc. Technol.14,(1977) 961
/13/ T. Hiramatsu, H. Goto, T. Hirobe, Y. Hirofuji and M. Kimata; J. Appl. Phys. 18, (1979), 853
/14/ T. Miyazaki, N. Nakamura, A. Doi and T. Tokuyama; Jpn. J. Appl. Phys. 13, Suppl. 2 pt2, (1974),441
/15/ J.A. Cooper Jr, E.R. Ward and R.J. Schwartz; Solid State Electron.15, (1972), 1219
/16/ J.E. Foster and J.M. Swartz; J. Electrochem. Soc. 117, (1970), 1410
/17/ L.G. Meiners,; J. Vac. Sci. Technol., 15 2-1407, (1978), 140
/18/ H. Nishi, S. Okamura, T. Inada, T. Katoda, H. Hashimoto and T. Nakaura; Inst. Phys. Conf. Ser. No.63 Chap.8,(1981),365
/19/ S. Okamura, H. Nishi, T. Inada and H. Hashimoto; Appl. Phys. Lett. 40 (1982) 689
/20/ L.A. Messick; J. Appl. Phys. 47 (1976), 5474
/21/ W. Quast; Electron. Lett. 8 (1972), 419
/22/ K. Tanaka, H. Takahashi, S. Kunikoahi and T. Ohki; Solid-State Electron.23, (1980), 1093
/23/ K. Ploog, A. Fischer, R. Trommer, and M. Hirose; J. Vac. Sci. Technol. 16, (1979), 290

/24/ G.D. Bogratshvili, R.B. Dzhanelidze, N.I. Pashintsev, O.V. Saksagarski and V.A. Skorikov. Thin Solid Films; 56, (1979),209
/25/ H. Nishi and A.G Revesz; J. Vac. Sci. Technol. 16, (1979), 1487
/26/ R.K. Smeltzer and C.C. Chen; Thin Solid Films;56, (1978), 75
/27/ I. Shiota, N. Miyamoto and J. Nishizawa; Surf. Sci.;86 (1979), 252
/28/ T. Hariu, T. Usuba, H. Adachi and Y. Shibata; Appl. Phys. Lett.; 32 (1978), 252
/29/ H. Beche, R. Hall and J. White; Solid-state Electron.; 8 (1965), 813
/30/ H. Hasegawa and T. Sawada; Thin Solid Films; 103, (1983), 119
/31/ C.C. Chang, R.P.H. Chang and S.P. Muarka; J. Electrochem. Soc. 125, (1978), 481
/32/ G.P Schwartz, B. Schwartz, D.D. Stefano, G-J. Guarotoero and J.E. Griffiths; Appl. Phys. Lett.; 34, (1979), 205
/33/ P.A Breeze, H.L. Hartnagel and M.A. Sherwood; J. Electrochem. Soc. 127, (1980), 454
/34/ G. Lucovsky and R. S. Bauer; J. Vac. Sci. Technol. 17,(1980), 946
/35/ W.E. Spicer, P.W. Chye, P.R. Skeath, C-Y.Su and I. Lindon; J. Vac. Sci. Technol. 16(5), Sep/Oct (1979), 1422
/36/ H. Hasegawa and T. Sawada;Thin Solid Films, 103 (1983), 119
/37/ H. Hasegawa and H. Ohno; J. Vac. Sci. Technol. 34, July/Aug (1986)
/38/ N. Yokoyama, T. Mimura, M. Fukuta and H. Ishikawa; ISSCC Dig.-Tech. Paper. Feb. (1981), 218
/39/ Y. Nakayama, K. Suyama, H. Shimizu, N. Yokoyama, H. Ohnishi, A. Shibatomi and H. Ishikawa; IEEE J. Solid-State Circuit, SC-18, 5, Oct. (1983), 599
/40/ N. Yokoyama, H. Onodera, T. Shinoki, H. Ohnishi and H. Nishi; IEEE, Trans. E.D. , ED-32, 9, Sep. (1985), 797
/41/ H. Nishi, S. Okamura, T. Inada, H. Hashimoto, T. Katoda, and T. Nakamura; Inst. Phys. Ser. No.63, Chap.8 (1981),365
/42/ T. Nakamura, A. Ushirokawa, and T. Katoda; Appl. Phys. Lett. 20-1,(1981), 223
/43/ J. Gyulai, J.W. Mayer, I.V. Mitchell and V. Ridorignev;Appl. Phys. Lett. 17 (1970), 332
/44/ B. Molnar,; J. Electrochem. Soc., 123, (1976), 767
/45/ H. Onodera, H. Kawata, N. yokoyama, H. Nishi and A. Shibatomi; Trans. E.D. , ED-31, 12, Dec., (1984), 1808
/46/ W.G. Oldham and S.S.Naik; Solid-State Electron, 15, (1972), 1085
/47/ M. Ozeki, K. Kodama, M. Takikawa, and A. Shibatomi; J. Vac. Sci. Technol. 21, 2 July/ Aug/ (1982), 438
/48/ K. Kitahara, K. Nakai, A. Shibatomi and S. Ohkawa; Jpn. J. Appl. Phys. 19, 7 July (1980), L369
/49/ T. Ohnishi, Y. Yamaguchi, T. Onodera, N. Yokoyama and H. Nishi; in Ext. Abs. 1984 Int. Conf. Solid State Devices and Materials (Kobe Japan) ,(1984), 391
/50/ P.M. Asbeck, C.P. Lee and M.F. Chang; IEEE Trans. E.D. ED-31, (1984), 1377
/51/ M.F. Chang, C.P. Lee, P.M. Asbeck, R.P Vahrenkamp and C.G. Kierkpatrick; Appl.Phys. Lett. 45, (1984), 279
/52/ T. Ohnishi, T. Onodera, N.Yokoyama and H.Nishi; IEEE, E.D.L., EDL-6, (1985)172
/53/ T. Onodera, T. Ohnishi, N. Yokoyama, and H. Nishi; IEEE Trans. E.D. ED-32, 11, Nov. (1985), 2314

Author Index